The Mobile Media Reader

Steve Jones
General Editor

Vol. 73

The Digital Formations series is part of the Peter Lang Media and Communication list.
Every volume is peer reviewed and meets
the highest quality standards for content and production.

PETER LANG
New York • Washington, D.C./Baltimore • Bern
Frankfurt • Berlin • Brussels • Vienna • Oxford

The Mobile Media Reader

EDITED BY Noah Arceneaux & Anandam Kavoori

PETER LANG
New York • Washington, D.C./Baltimore • Bern
Frankfurt • Berlin • Brussels • Vienna • Oxford

Library of Congress Cataloging-in-Publication Data

The mobile media reader / edited by
Noah Arceneaux, Anandam Kavoori.
p. cm. — (Digital formations; v. 73)
Includes bibliographical references and index.
1. Telecommunication—Social aspects. 2. Communication and culture.
3. Mobile computing. 4. Multimedia systems.
I. Kavoori, Anandam P. II. Arceneaux, Noah.
HM851.M628   384—dc23   2011033865
ISBN 978-1-4331-1301-7 (hardcover)
ISBN 978-1-4331-1300-0 (paperback)
ISBN 978-1-4539-0212-7 (e-book)
ISSN 1526-3169

Bibliographic information published by **Die Deutsche Nationalbibliothek**.
Die Deutsche Nationalbibliothek lists this publication in the "Deutsche
Nationalbibliografie"; detailed bibliographic data is available
on the Internet at http://dnb.d-nb.de/.

Contents

Introduction: Mapping Mobile Media   1
Noah Arceneaux & Anandam Kavoori

Foundations

1. Historicizing Mobile Media: Locating the Transformations   9
    of Embodied Space
    Jason Farman

2. Calling Ahead: Cinematic Imaginations of Mobile Media's   23
    Critical Affordances
    Scott W. Ruston

3. Analog Analogue: U.S. Automotive Radio as Mobile Medium   40
    Matthew A. Killmeier

4. CB Radio: Mobile Social Networking in the 1970s   55
    Noah Arceneaux

5. A Brief History of U.S. Mobile Spectrum   69
    Thomas W. Hazlett

Forms/Functions

6. Not TV, Not the Web: Mobile Video Between Openness and Control   87
   Aymar Jean Christian

7. Reading After the Phone: E-readers and Mobile Media   102
   Gerard Goggin and Caroline Hamilton

8. As It Happens: Mobile Communications Technology, Journalists   120
   and Breaking News
   Collette Snowden

9. Time and Space in Play: Saving and Pausing with the Nintendo DS   135
   Samuel Tobin

10. You Can Ring My Bell and Tap My Phone: Mobile Music, the Ringtone   148
    Economy, and the Rise of Apps
    Ben Aslinger

11. Appropriation of Cell Phones by Kurds: The Social Practice of Struggle   163
    for Political Identities in Turkey
    Burçe Çelik

12. Through the Looking Cell Phone Screen: Dreams of Omniscience   177
    in an Age of Mobile Augmented Reality
    Imar de Vries

Contributors   191

Index   195

Acknowledgment

The original version of the material that appears here as Chapter 9, "Time and Space in Play," can be found in:

Samuel Tobin, "Time and Space in Play: Saving and Pausing with the Nintendo DS," *Games and Culture*, by SAGE Publications, Inc. (Publication forthcoming). Reprinted by Permission of SAGE Publications, Inc.

Introduction:
Mapping Mobile Media

NOAH ARCENEAUX & ANANDAM KAVOORI

In all these years that I've been carrying it and reading it every day, I got so caught up
in keeping it safe that I forgot to live by what I learnt from it.
— *THE BOOK OF ELI* (WARNER BROTHERS, 2010)

Denzel Washington—who plays the central character in *The Book of Eli*—is a
study in mobility—physically moving with assured grit through a post-apocalyp-
tic terrain, driven in equal part by theological certainty and human will. Politically,
the media he carries (a Bible) can move nations and people. As the villain (Carnegie)
in the film puts it, "It's not a book, it's a weapon...people will come from all over,
they'll do exactly what I tell them if the words are from the book...all we need is
the book." But above all, it is the book's religious mobility—its ability to create and
carry a portable liturgy, a religious sensibility and indeed a fully assembled world
view that is the *leitmotif* of the entire film—and of Eli's life journey.

We begin with this (extended) popular culture reference to suggest that while
some specific forms of mobile media may be "new," the general concept itself is not.
Mobile media are at least as old as the surfaces (stone, papyrus, metal) that allowed
a story to be carried, and begins even earlier with oral culture—a system of mobility that, for example, allowed the *Rigveda* (a Hindu epic) to be passed on by word
of mouth for at least two thousand years before it was formalized into text. This is
also the intent and scope of this volume, which has a number of threads running

through it—the most important of which is a much-needed historicization of mobile media.

The current hype and hyperbole associated with mobile technologies threatens to overwhelm serious insight into the phenomenon, as relentless advertisements tout the lightning-fast speeds of various networks, the benefits of staying in touch with loved ones, and (somewhat comically) the ability to watch wide-screen, high-definition movies on miniaturized screens. The commercial media industries have seized upon this platform, revamping existing content for the particular demands of phones or creating entirely new forms. While many media industries established long ago suffer from declining audiences and dwindling advertising rates (most notably print media), there is something of a mania for "apps," and mobile media, which derives from all previous forms, shows potential for even more significant growth. Such aggressively optimistic and even utopian hopes recall similar claims made about earlier electronic technologies, dating back to the telegraph. Without entirely rejecting such claims, and without neglecting some of the bona fide benefits of portable phones, this collection encourages readers to evaluate the phenomenon of mobile media within a broader, critical context.

This book had its beginnings in *The Cell Phone Reader: Essays in Social Transformation*, a 2006 anthology that sought to theorize a new medium. The earlier book brought together essays that theorized cell phones in a variety of ways—as a symbolic network, as an artifact of popular culture, as an agent of the public sphere and so forth—along with an understanding of the cell phone across a range of contexts and media/national cultures. This edited collection draws on some of the themes of *The Cell Phone Reader* but is made up of an entirely new set of essays—and for the most part by new contributors.

Given the dramatic changes associated with mobile technology in the intervening years, we realized that the earlier collection could benefit from an update. At the time of the original inception of *The Cell Phone Reader*, the academic study of cell phones was in a relatively early stage, though subsequent years have seen no shortage of monographs, anthologies, and journal articles that analyze the social and cultural implications of these devices. Technological innovation has proceeded at perhaps an even more rapid rate, and as previously noted, these devices are no longer limited to voice and text communication and are quite literally portable, multi-media centers that fit neatly into one's pocket.

While many of the observations and insights from *The Cell Phone Reader* remain relevant, so many features have been incorporated into phones that the word "phone" itself feels quaint and outdated. As we contemplated the various issues and concerns raised by the growth of mobile media, we realized that an entirely new anthology was in order. Due to the real-world parameters of the publishing process, there will likely be significant developments related to mobile media that will occur in the period between the writing of these essays and the final publication. As with our earlier collection, though, we believe that the essays will remain relevant for years to come, and offer enduring insights.

This new anthology is justified not only because of technological developments, but because the varied uses of mobile communication devices present an array of topics for intellectual interrogation. Working from the premise that technologies are inherently cultural objects, and not just combinations of micro-circuitry and cutting-edge software, the essays in this anthology explore the negotiations between industries that inform and shape the content that is delivered to mobile devices. Another recurring theme is the importance of cultural perceptions; the ways in which new technologies are conceptualized and imagined can influence their evolution just as much as anything from the realms of technology or engineering. These cultural critiques are infused with empirical, real-world evidence, thus providing readers with information about what has happened in the past and the current status of various technologies. Our hope is that along with factual knowledge about mobile media, readers will find suggestions as to how to comprehend and understand this latest evolution in communications technology.

We have been guided by three maxims in our choice of essays—*historicize, contextualize* and *diversify*. The first refers to understanding mobile media in both its historical context and historicizing contemporary mobile media; the second to finding frames of reference that draw on a number of intellectual traditions, so that the study of mobile media itself matures—and finally, diversify refers to the range of topic areas (outside the usual ambit of technology or usage) that a complete study of mobile media should aspire to. In what follows, we will map some of the themes/subjects of the individual chapters as they relate to these maxims. Needless to add any one volume on this subject can scarcely do complete justice to the mass of technologies that makes up "mobile media" today. Our goal has been to cover as many topics as possible, drawing from a variety of academic approaches, though there are undoubtedly aspects of the phenomenon that we have omitted.

The first section of the book, "Foundations," provides, as the title suggests, a foundation for thinking about mobile media today. Taken together the five essays in this section historicize, contextualize and diversify our thinking about mobile media by taking a relatively circumscribed area of analysis (location-aware technologies; cinematic representations; postwar audio radio; citizens band radio; and spectrum allocation) and connecting them with wider institutional/technological and sociological elements in play.

Jason Farman ("Historicizing Mobile Media: Locating the Transformations of Embodied Space") explores the tensions around the growth of location-aware technologies by looking at earlier forms of mobile media. Early in his essay he sets the terms for understanding mobile media that apply to all the chapters that follow and are worth repeating in this introduction: "The historical blind spots in our contemporary imaginings of mobile media come out of a tradition of technological determinism…and while it is true that technologies shape and transform our lives to some extent, this perspective tends to ignore the cultural and historical forces that were responsible for the emergence of a particular technology." In that spirit, this book (like its predecessor) *centers technology as a cultural form*—or what Peter Brosius, an anthropologist at the University of Georgia, calls "technologies of visualization": media forms that *make concrete* the ceaseless flow of culture(s) in post-industrial society. Farman focuses on a common denominator of mobile media—how our uses of technologies transform our relationships to social space. Drawing on historical examples (papyrus, the printing press, clocks, and contemporary media), he suggests specific tensions between notions of proximity/intimacy and notions of distance/otherness that exist in the cultural transformations that accompany the expansion of mobile technologies.

Scott W. Ruston ("Calling Ahead: Cinematic Imaginations of Mobile Media's Critical Affordances") suggests that mobile media exhibits a unique combination of *affordances* which include "ubiquity"—the quality of being always in our presence, always in our possession, and always connected to a larger mediascape; "portability"—the ability to take a device with us wherever we go; "personality"—which encapsulates individuality and bodily materiality (the concept extends beyond the mere customization to an overall cyborgian quality of an extension of the self and body); "connectivity"—the technological and social sense of human interaction and engagement; and finally, "locativity"—the capacity of the mobile phone to be out in space, unbound from a fixed location, unlike the television and cinematic experience, and make a relevant association between media content and place for the user. He then examines cultural locations where such affordances can be observed and finds them in what he calls "the cultural imaginary"—the complex set of media images that come to us from cinema, television, and new media. Focusing on television advertising, Ruston provides a methodological template for thinking through the complex chains of media and cultural connectivity that make up the representation and use of mobile media.

Matthew Killmeier ("Analog Analogue: U.S. Automotive Radio as Mobile Medium") explores the history of postwar U.S. automotive radio. He identifies the unifying and segmenting social forces that shaped automotive radio along with dialectically intertwined centrifugal and centripetal tendencies. He offers a general theoretical template for understanding contemporary mobile media: they build upon existing communications linkages, but over time reconfigure them as new communications exchanges; successful mobile content complements the particular characteristics of the mobile medium's development and cultural usage; mobile media are characterized by mobile privatization, mutually constitutive with segmentation and increasing everyday and aggregate mobility; and finally, there is a homologous, mutually constitutive relationship between mobile media form and content. He argues that in order to comprehend form and content, we have to understand the medium as a cultural form—and further, while cultural form encompasses a particular medium with technological features, it extends *beyond* the medium and its political-economic contextual framework to privilege social practices and signification.

Noah Arceneaux ("CB Radio: Mobile Social Networking in the 1970s") takes a serious look at one of the most popular fads of that decade, citizens band radio. The technology had existed for some time, though a combination of factors catapulted this form of communication into the mainstream in the early 1970s and drivers of every persuasion, not simply truck drivers, added them to their dashboards. By analyzing popular representations of citizens band, including depictions in films and country songs, Arceneaux identifies some themes that continue to be associated with mobile communication. CB radio was, for example, seen as somewhat rebellious, unruly technology that allowed drivers to evade authorities; today, mobile communication is also seen as something that is difficult to control.

Thomas Hazlett's chapter ("A Brief History of U.S. Mobile Spectrum") is a departure from other essays in this collection in its relative narrow focus on the connections between technology, institutional developments, public policy prescriptives and marketplace dynamics in the development of what he calls the "mobile spectrum." While some savvy consumers are aware of the importance of spectrum allocation, many have no knowledge of this pivotal issue that underlies mobile communication. Decisions about which companies have access to the airwaves, and how much they might utilize, exert significant influence upon the services that are available. Hazlett provides a historical tracking of the various factors that determine spectrum allocation and makes a compelling argument: "The history of mobile services in the U.S. reflects a distinct pattern. Long delays block economic development; when the bottleneck loosens, permitting dollops of new bandwidth to go to entrepreneurs, frenetic economic activity ensues—new services arise, prices fall, and mobile use skyrockets." This dynamic conforms to the longer history of spectrum allocation, as engineers and technicians continually seek to maximize the number of messages that can be carried over particular frequencies.

The second section of the book, "Forms/Functions," reiterates the broad themes of the book—historicize, contextualize and diversify—through seven essays that address specific forms of mobile content and ways in which they are utilized: mobile video, the growth of e-readers, journalism, gaming, ringtones, ethnic identity and augmented reality. What connects these chapters is their critical intent—rethinking an area of mobile media, fashioning a language in which they might be interrogated and offering a direction for future studies of the forms and functions that make up mobile media.

Aymar Jean Christian ("Not TV, Not the Web: Mobile Video Between Openness and Control") offers a critique of scholars who see the web as open and mobile as closed (through market and technology) by examining the efforts of three distributors of independent web video—Vimeo, My Damn Channel, and Q3030 Networks—alongside larger video sites—YouTube, Hulu and Crackle—to show how navigating the mobile market involves negotiating industrial and technological considerations, sometimes open and closed, sometimes neither. He suggests that the mobile medium remains at once highly volatile, deeply complex and somewhat confusing for video distributors, who must navigate numerous layers of the industry to reach users. This complexity, he suggests, makes mobile media a space of compromise, possibility and challenge.

Gerard Goggin and Caroline Hamilton ("Reading After the Phone: E-readers and Mobile Media") make a central claim—reading has long been a portable media form—and then go on to map and theorize "reading" as an element of contemporary mobility, focusing on the e-reader. Beginning with the rise of the cell phone and the "texts" it spawned, they proceed to give an overview of the emergence of the e-reader in the field of mobile media and when and how they begin to figure in mobile devices. They discuss the early uses of mobiles for reading that largely developed independently of e-reader technology plans. They then move on to discuss the centrality of e-readers as an integral part of mobile media in the smartphone era. This is something that they argue helps us broaden our ideas of what mobile communication media is, and complicates the politics, design, and direction of this eminently pervasive technology.

Collette Snowden ("As It Happens: Mobile Communications Technology, Journalists and Breaking News") takes a close look at the impact of mobile communication technologies on journalists' routines and practices. Focusing on live, breaking news she argues that the use of mobile communications technology is now so widespread and commonplace that it has disrupted the conventions governing the public distribution of news and information. The work of journalism has been radically redefined by the blurring of responsibility for the production of content between journalists and audiences, particularly in breaking news situations, especially where events are unplanned, unexpected, in hard-to-access locations, or on a large scale. She goes on to examine some of the concepts/practices that have emerged in this historical moment, focusing especially on the concept of "storify"—how the amount of user-generated content is being managed and how journalistic processes and practices are being modified in response to the amount of content available. The focus on the aggregation of content and its contextualization, she argues, is not a new journalistic practice, but rather the extension and intensification of an existing one, using new technological platforms. Snowden's work is an important example of exploring the emergent language around journalism as the news media move to a greater emphasis on their role as aggregators of information rather than originators.

Samuel Tobin's essay ("Time and Space in Play") is focused on the Nintendo DS, an enormously popular mobile technology that has been relatively understudied, as a way of talking about the broader phenomenon of mobile gaming. Using Goffman's and Benjamin's theories of play and Innis's analysis of media, he explores the nature of play on a handheld video game system as an illuminating case of the tension between time-based storage media and space-based transmission media. Storage, transmission, space, and time are intertwined and made complicit in the ways in which the Nintendo DS is used and played. He argues that by engaging with non-diegetic aspects of the video game experience, such as saving and pausing, we can begin to address the materiality of handheld video gaming systems as objects with which we play with and rework time and space. In turn, he argues, mobile play itself is highly contingent and spatial-temporal practices take on special significance in this light.

Ben Aslinger's essay ("You Can Ring My Bell and Tap My Phone: Mobile Music, the Ringtone Economy, and the Rise of Apps") theorizes the "mobile music soundscape" as a complex shifting site of meaning making shaped by "struggles over power within the recording and music industries between artists, publishers, performance rights organizations, major labels, and 'indie' labels; the emergence and decline of aggregators; the number of wireless carriers operating at the multinational, regional, national, or local level...; and conflicts between carriers, aggregators, labels, and publishers over who controls the mobile music market, modes of distribution, and who 'owns' the consumer." He then proceeds to map out the terrain towards a full accounting of the mobile music soundscape by charting the evolution of the ringtone market, mobile music genres and the question of "quality," along with the way in which mobile music has intersected with public anxieties about its omnipresence. He concludes with recent developments, focusing on the exploding use of a variety of apps as they relate to mobile music.

Burçe Çelik ("Appropriation of Cell Phones by Kurds: The Social Practice of Struggle for Political Identities in Turkey") explores the role of cell phones within the political and cultural struggle of young Kurds in Turkey, a country whose Kurdish citizens, who make up more than 10 percent of its total population, have been denied their cultural rights. Although this essay is about one particular ethnic group, the Kurdish situation is sadly not unique, and it is hoped that readers everywhere can find something to relate to. In particular, Çelik focuses on how the cell phone, through its containing space for textual, visual and sonic content, is integrated into everyday politics of young adult Kurds as a means of claiming their cultural identity, asserting their mother language as a language of technology and modernity. Based on in-depth interviews, Çelik provides one of the few studies of texting as a political/cultural practice. Texting, she argues, "becomes a way for the Kurdish users to practice their language—not because it is a fast, easy or cheap way of connected presence; on the contrary it is slow, time-consuming and as cheap as talking on the phone. Yet, texting allows them to improvise, perform and experiment with their mother language, which remains unfamiliar to most of them in its written form. The use of improvised written Kurdish in SMS becomes a performative and dutiful practice that is assumed to convey the idealized image of self in the most desired language, which has been subjected to severe interdiction."

Finally, Imar de Vries's essay ("Through the Looking Cell Phone Screen: Dreams of Omniscience in an Age of Mobile Augmented Reality") is a theoretical accounting of an Amsterdam-based augmented reality project along with a critical analysis of the myth of constant connectivity and omniscience. He charts the journey of "the dream of omniscience" through industry, technology and popular culture discourse, suggesting that mobile media are an expression of a "technological imaginary" that holds that radically heightened connectiveness will bring us closer to a utopia of omniscience, omnipresence and complete understanding. He argues that the other side of joining this mobile revolution is that users are forced to open up the logic of the network in which every node is part of a ubiquitous web of information.

Each of these chapters cut across the three principles we have used for this collection—they *historicize* a specific realm of mobile media (music, cinema, video, amongst others), *contextualize* it outside of a narrow technological or institutional context and *center it as a cultural text* (including theoretical essays on embodied space, augmented reality and critical affordance). Finally, they *diversify* the scope and range of mobile media studies, bringing scholarly attention to areas that have been seldom analyzed by media scholars (including e-readers, Nintendo DS, citizens band radio, auto radio and ringtones).

Needless to add, the import and impact of this *Mobile Media Reader* lies in the next step in its textual journey—with you, the reader.

1

Historicizing Mobile Media

Locating the Transformations of Embodied Space

Jason Farman

On a morning in late May of 2010, I woke up to the sound of my iPhone alarm clock, went downstairs for a cup of coffee, and read the morning news through an app I had downloaded the day before. About halfway through my coffee, I came across a story with the headline "Foxconn Suicides Continue: Inside the iPhone Factory Plagued by Deaths." The story noted that there had been 11 suicides since the beginning of the year in the Chinese factory that was responsible for the production of the majority of iPhone, iPad, and iPod units sold worldwide. I stopped my morning ritual to read about the working conditions in the plant, which included long hours of mandatory overtime and rigid management on the factory floor that enforced an incredibly fast assembly line. Just hours after the factory's chairman gave reporters a tour of the facility, noting that the company would "work harder to prevent more deaths," the 11th worker jumped to his death.[1] At that moment, it dawned on me: I had never wondered where my mobile phone came from. I knew about its design and I knew that it contained elements gathered from around the globe, but it was all very vague to me. This moment is what motivated me to look into the pathways (both historical and global) that led to my mobile phone.

As I sat in my home in the United States reading about the deaths of workers in China on the very device that they helped create, I realized that within this one scene (in which I played the role of the consumer) are multiple histories of mobile media, their place within culture, and their role as a cultural force. By tracing the various histories of mobile media, I hoped to get a clearer picture of the pathways that led to my phone. In doing so, I discovered many important transitional moments throughout history in which a medium that was once considered geographically fixed or static becomes mobile. These moments are almost always situated among major cultural shifts. Part of the process of historicizing mobile media is to connect these moments and shifts to the cultural landscape and "old" media that led to the technologies and imaginaries of mobility. Emerging alongside these cultural shifts is a changing relationship between notions of intimacy and distance as they are attached to concepts of the local and the foreign. As I argue in this chapter, as various mobile media are created, their inherent social and spatial characteristics cause a continued interrogation of the relationships between proximity/ intimacy and between distance/otherness.

An investigation into these cultural shifts necessitates a broad definition of mobile media. Currently, mobile media tend to be grouped into a larger category of emerging digital technologies, often termed new media. However, the notion that mobile technology is "new" is indeed shortsighted since, as we will see throughout this chapter, mobile technologies have been around in some form throughout human history. The historical blind spots in our contemporary imaginings of mobile media come out of a tradition of technological determinism, characterized by the notion that once a "new technology is introduced it reformulates society in its image."[2] From a technological determinist perspective, media often storm the cultural landscape, appearing because they were the next logical step along a pathway to progress that is unstoppable. We always want our technologies to improve and to transform the ways we live our everyday lives. And while it is true that technologies shape and transform our lives to some extent, this perspective tends to ignore the cultural and historical forces that were responsible for the emergence of a particular technology. Thus, by historicizing mobile media, we can trace the rich genealogical roots that served as the foundation for the media you are currently reading this on, whether it be a print-bound book, an e-reader, or a laptop.

Before engaging in this type of historical inquiry, it is worth echoing Lisa Gitelman's concerns about what exactly it is that we are historicizing when tracing the roots of new media. She asks, "Is the history of media first and foremost the history of technological methods and devices? Or is the history of media better understood as the story of modern ideas of communication? Or is it about modes and habits of perception? Or about political choices and structures?"[3] In response to her own questions, she offers a cautionary word of advice for historians of media. For Gitelman, it is vital to historicize with the understanding that history is weighted down with "a host of assumptions about what is important and what isn't—about who is significant and who isn't," and these assumptions also echo into the various meanings of media. She writes, "If there is a prevailing mode in general circulation today, I think it is a tendency to naturalize or essentialize media—in short, to cede to them a history that is more powerfully theirs than ours."[4] Therefore, the act of tracing out the histories of mobile media requires us to be mindful that we aren't simply discussing how one technology eventually conceived of a newer, more robust version that changed culture in its wake. Instead, we are positioning mobile media within the cultures that create them, locating the moments that cleared the way for these media to take hold and become dominant media forms for communication and art. Simultaneously, we are looking at, as N. Katherine Hayles puts it, the ways that culture and the people shaping culture "can and [do] transform in relation to environmental selective pressures, particularly through interactions with technology."[5]

Histories of Mobile Media and Reconfigurations of Social Space

One common denominator among mobile media is the way our uses of these technologies transform our relationships to social space. Contemporary experiences with the Internet show one example of this transformation: while we have for decades connected to the Internet from stationary computers, with the move to mobile computing on devices like smartphones, we are able to link data to location in unique ways that give us a renewed sense of how social space is produced. While such an experience of using the Internet on the go is now commonplace, the affects (and artifacts) of this shift in usage should not be underestimated. As I will develop later in this chapter, this emergence of mobility through mobile computing devices creates a paradigm shift similar to the cultural changes brought about by mobile writing tools. Take papyrus as an example: being one of the earliest forms of mobile media, papyrus was a paper-like medium made up of thin cuts of the papyrus plant bonded together and used for writing in Egypt as early as the third millennium B.C. Here, the mobility of the medium worked in conjunction with the mode of writing it used. Moving from stone inscriptions, which required a great deal of time to complete, the use of brushes on papyrus allowed the mode of writing to more closely match the speed of thought. This was especially true, as Harold Innis notes, as the Egyptians' use of hieroglyphic writing transformed into script that better matched the writing instruments used (brushes instead of chisels).[6] This afforded more flexibility and experimentation in what could be inscribed, since the permanence and amount of time required for the inscription were much less significant than writing on stone. This speed of inscription matched the speed and distance of transmission: written ideas were no longer geographically bound to a specific location but could be transported across vast distances. People no longer had to travel to a site to read the writings inscribed on caves, monuments, or walls; instead, the ideas traveled broadly since the medium they were inscribed on was light and mobile. These changes are remarkably similar to the ones we're experiencing with mobile computing devices: we are seeing a rapid increase in the speed and frequency of communication as well as an alteration in our conceptions of social space.

This type of mobility has worked in conjunction with many political and cultural shifts throughout history. It's difficult to think about the mobility of ideas without thinking of the powerful cultural transformations brought about by the uses of the printing press and how it was used to launch the Christian Reformation. The printing press, invented by Johannes Gutenberg in 1440, created mobile media such as pamphlets and books and the press itself was transported to different cities and set up to make business with local businesses (such as churches) and citizens.[7] The most revolutionary aspect of these mobile media was their ability to transform social spaces. The cultural shift that happened in conjunction with the printing press can be mapped onto our uses of mobile media (especially location-aware technologies): the cultural imaginaries of space became simultaneously about experiencing the expansion of space, an increase in speed of transmission, and a transformed view of the local. Space expanded, in part, because ideas were spreading across distant regions and it became very difficult to remain in a space that was unaffected by distant cultures. Extending the speed of writing that the papyrus, brush, and ink offered, the printing press allowed for the reproduction of ideas to move at an unparalleled pace, as a massive number of copies could be made from one typeset press. As these ideas spread rapidly, cross-cultural influences affected the space of everyday life, changing common practices as alternative forms emerged, and also forcing a reevaluation of the local practices and beliefs that had been engrained into local traditions.

These cultural shifts came about because of the uses of mobile media. Tracing similar relationships between mobile media and cultural shifts in knowledge and power can be easily accomplished. For example, we can trace this familiar pattern onto the remediation of the clock tower to the home clock, the pocket watch, and eventually the wristwatch.[8] Each of these later remediations are powerful mobile media that have, in one way or another, transformed relationships to social space. The clock has been described as "an instrument of social control," which was seen as "supplanting nature and God with clocks and watches...[and] with secular authorities based on efficiency and convenience."[9] The sense of expanding space and its impact on local space, as discussed above, was at the core of the adoption of this technology. As the world became more mobile in the 19th century with railroads and in conjunction with communications across vast distances through the telegraph, people had to continually adjust their watches to account for local time. This meant that as someone traveled east or west across the United Stated on a train, this person had to adjust his or her watch by one minute for every 12 miles they passed. This was common in the mid-1800s until standardized time zones were created in the United States in 1883 and international time zones in 1884.[10] Thus, after 1884, people carrying around a pocket watch were aware that the notion of "local time"—and thus what constituted local space—had changed dramatically. A watch with the "correct" time was set to international standards, based on Greenwich Mean Time in England. This mobile technology connected the individual to a sense of global space and time, while changing what it meant to travel and live in local space. The city of Detroit, for example, refused to comply with standardized time zones for many years since it was geographically located roughly halfway between the Eastern and Central time zones. To comply with these international standards meant that the citizens of Detroit would have to dramatically change their understanding of local time and space. For those who lived close to the borders between time zones, traveling just a few short miles could move time forward or backward by an hour, causing people who are spatially close to one another to live in different notions of "local" time.

As evidenced by these mobile technologies, there is an interesting tension between notions of proximity/intimacy and notions of distance/otherness in these cultural transformations that accompany the expansion of mobile technologies. This tension is seen in the mobile media players/listening devices that have been utilized in recent history. The remediation of personal music devices (such as the iPod or any generic form of digital music and media player) comes from, as Michael Bull notes, "previous generations of mobile music reception."[11] He goes on to argue, "Mobile sound technologies and their use do not exist in a cultural vacuum—prior to the Apple iPod came the personal stereo. Prior to that, the transistor radio and the portable record player and of course there exists a history of mobile listening in automobiles through radios and then cassette players."[12] Throughout the history of mobile sound technologies, their usage has either functioned to connect people (especially across distances, as seen with the transistor radio or CB) but also to "cocoon" people from those in their immediate vicinity. Mizuko Ito has argued that the personal listening device can serve to help isolate people from social situations, to "cocoon" them from the need to interact with others in a crowded environment. Ito, along with Daisuke Okabe and Ken Anderson, writes, "Cocoons are micro-places built through private, individually controlled infrastructures, temporarily appropriating public space for personal use....These cocoons also have specific temporal features, functioning as mechanisms for 'filling' or 'killing' in-between time when people are inhabiting or moving through places within where they are not interested in fully engaging."[13] Thus, mobile listening devices have served to help people fill "in-between time" by giving them an activity that removes them from the need for interacting with the people around them. The tension here exists in how a person exists in and among a crowd of people (in their intimate proximity) but is also removed from them (by making him or herself distant through mobile media).

The inverse of this spatial tension takes place with mobile technologies such as personal radios and citizens band (CB) radios: though people are geographically distant from one another, they experience a sense of community and intimacy through the use of these technologies. At a recent conference, a professor noted to me that while he had never used the mobile social media I was discussing—location-aware social networks such as the early examples of Foursquare, Gowalla, and Loopt—he immediately understood the concept of connecting with others in a spatial way through mobile media. He understood the idea of location-aware social media through his use of the CB while he made a cross-country drive when he was younger. He bought the CB especially for the trip and had never used one before getting in the car to make the journey. As soon as he embarked, he began a series of dialogues with people, often beginning by getting the person's name, their location, and their destination. In this way, the travelers connected on the common themes of the journey, the landscapes, and the other drivers they encountered. The professor telling me the story remarked, "There were long stretches where I was driving alone on these vast, empty prairies. In any other situation, I would have felt extremely isolated from humanity. But by using the CB, I felt like I had companions with me as I drove." This sense of social connection across distant geographies, gained through utilizing mobile media, is what Clive Thompson has termed "social proprioception."[14] Here, proprioception is understood as more than just "your body's ability to know where your limbs are. That subliminal sense of orientation is crucial for coordination: It keeps you from accidentally bumping into objects, and it makes possible amazing feats of balance and dexterity," as Thompson puts it. It is a sense of your body in its place, which always orients itself in relational space. You position yourself (both as a body and as a subject) though spatial relationships to others and to objects. Essentially, you know where your body ends and the social world begins and can chart your body through space, navigating to important locations such as where you left your car or how to get from your house to your classroom. In a sense, all proprioception is social. However, Thompson's term gets to the heart of a unique feature of these technologies: we understand our space differently because of the social connections we make and these connections are increasingly incorporating a notion of the social that happens across vast geographic distances. The story offered by this professor who utilized CB radio is an especially apt one, since the mobile phones we use in very similar ways were dependent on the advancements made in radio communications.

The History of the Mobile Phone

Our contemporary uses of mobile phones might better resemble the CB radio communications of travelers than what the first mobile phone actually looked like. Built in 1910 by a Swedish electrical engineer by the name of Lars Magnus Ericsson, the first mobile phone was built into his wife's car. Ericsson, who retired early to a life on a farm after famously founding the telecommunications company named after him, would make journeys away from the farm and sought to find a way to stay in contact while he was on the move. John Meurling and Richard Jeans describe this early version of a mobile car phone:

> In today's terminology, the system was an early "telepoint" application: you could make telephone calls from the car. Access was not by radio, of course—instead there were two long sticks, like fishing rods, handled by Hilda [his wife]. She would hook them over a pair of telephone wires, seeking a pair that were free....When they were found, Lars Magnus would crank the dynamo handle of the telephone, which produced a signal to an operator in the nearest exchange.[15]

Though technically serving as the first "mobile" phone in the strictest sense, this concept was never utilized by the LM Ericsson Telephone Company and never served as the technological origins for modern wireless mobile phones.

The mobile phone as we know it was founded on technologies that sought to bring together the invention of the landline telephone in the 19th century and the invention of the radio around the turn of the 20th century. As radio became increasingly popular in the 20th century, people were able to have two-way communication while on the move in very similar ways to our voice communication on a mobile phone; yet the limits of the technologies at the time hindered this radio communication in many ways. One of the first problems faced was the conflict between the power source and the portability of radio. Early radios required a massive amount of power and thus required an equally massive battery. This limited the use of radio to naval ships initially, and automobiles later on, which were each large enough to house the equipment and the battery required to power the device. Here, there is a significant distinction between *portable* and *transportable*. These early radios were transportable, but far from serving as a portable/mobile technology as we understand it. Jon Agar notes,

> One of the most important factors allowing phones to be carried in pockets and bags has been remarkable advances in battery technology. As batteries have become more powerful, so they have also become smaller. Partly because improvements in battery design have been incremental, their role in technological change is often underestimated.[16]

Here, there is an indelible connection between the power source and the portability of the device.

A second limitation faced by radio, as it informs its eventual remediation onto mobile phones, was the scarcity of radio frequencies. Radio waves traveled across limited frequencies and, as more users populated those frequencies with their conversations, the radio waves became crowded with overlapping conversations, eventually making them unintelligible. As Rich Ling and Jonathan Donner point out, "In the first radio telephone systems for New York City, only a handful of users were able to simultaneously make a call. In that system there was only a single radio cell for a whole city."[17] To address this issue for our current mobile phone communications technologies, engineers at Bell Laboratories came up with the concept of the "cellular idea" in 1947. This concept, which gives us the term "cell phone," allowed for true mobility with early radio telephones without interference from other callers. Agar describes the "cellular idea" in detail:

> Imagine a map of New York City and imagine a clear plastic sheet, ruled with a grid of hexagons, placed over it. Now, imagine a car, equipped with a radio telephone, driving through the streets, passing from hexagon to hexagon....If each hexagon, too, had a radio transmitter and receiver, then the radio telephone in the car could correspond with this "base station." The trick... was to allocate, say, seven frequencies to a pattern of seven hexagons (a–g), and repeat this pattern across the map. The driver started by speaking on frequency a in the first hexagon, then with g, then c, and then back to a again. If the first and last hexagon were far enough apart so that the two did not interfere... then a radio conversation could take place without interference....Suddenly, if the hexagons were made small enough, many more mobile telephones could be crammed into a busy city, and only a few scarce resources would be needed.[18]

But, in 1947, the technology was not prepared to handle this rapid mobility, in which the handoff between cells needed to be so rapid that the user didn't notice (and the user could be sufficiently tracked so that the handoff—and the payment—could be monitored by the telecommunications infrastructure). Thus, the advancements of the mobile phone needed to be backed by computing technologies that could support these rapid handoffs between cell towers and also required an obvious demand from the public to create a system that moved beyond radio conversations to devices that could communicate directly to a landline telephone.

When the technologies and the demand were both in place, the first cellular phone system was installed around Chicago in 1977. A few years later, in 1981, the system had 2,000 subscribers, which was all it could handle.[19] In 1980, trials of a cellular system were built in the Baltimore/Washington, D.C. area, and the demand for car phones became significant. As Agar notes, Motorola anticipated this demand and filed patents for cellular technologies as early as 1973.[20] Toward the end of 1983, Motorola released its first handheld mobile phone in the United States. Called the 8000x (and sometimes referred to as the "brick" phone due to its shape). Beyond the size and the exorbitant costs of the phone, the cellular technology it connected to was better suited to car phones, and the public's perception of the mobile phone initially pigeonholed it as a technology designed for the automobile. This last concern was one particularly centered in the United States, since "the entire US way of life revolved around the automobile."[21]

Mobile Device and Asynchronous Communication

Mobile phones have evolved significantly since the brick phone of the 1980s—both in shape, technological specifications, but perhaps more noticeably in the way that they are used. Howard Rheingold begins his book, *Smart Mobs,* with an observation he had in Japan that changed the way he thought about the uses of a mobile phone: "The first signs of the next shift began to reveal themselves to me on a spring afternoon. That was when I began to notice people on the streets of Tokyo staring at their mobile phones instead of talking to them."[22] This trend of looking at the phone (as a computing or texting device) instead of talking into the phone (as a voice communication tool) hit its nexus in 2009. In this year, for the first time ever, people used their mobile phones more for data transfer (web usage, mobile applications, and text messages) than for voice communication.[23] As a result, we are connecting with each other differently and also connecting with our spaces differently. As mobile media users continue to demonstrate a preference for asynchronous communication through either sending text messages or emails in lieu of making phone calls, these choices are coming under the age-old critique that media create social and spatial distance. Asynchronous communication (i.e., the creation of documents such as text messages instead of real-time interactions) is often seen as removing people from the immediacy of voice communication.[24] As Kenneth J. Gergen argues about forms of mobile communication that prioritize "absent presence" rather than synchronous engagement, "The erosion of face-to-face community, a coherent and centered sense of self, moral bearings, depth of relationship, and the uprooting of meaning from material context: such are the repercussions of absent presence. Such are the results of the development and proliferation of our major communication technologies of the past century."[25]

As I note in my book, *Mobile Interface Theory: Embodied Space and Locative Media,* the nostalgia in this sentiment is one which has accompanied all emerging forms of media, especially the advances in mobile media which often transmit various kinds of documentation.[26] By historicizing our mobile media, we see that such critiques of presence versus absence accompanied even the telegraph, which was criticized by Henry David Thoreau as that which will "distract our attention from serious things. They are but improved means to an unimproved end, an end which it was already but too easy to arrive at; as railroads lead to Boston or New York. We are in great haste to construct a magnetic telegraph from Main to Texas; but Main and Texas, it may be, have nothing important to communicate."[27] The landline telephone was also criticized as that which would distance neighbors from each other since there would no longer be a reason to visit when someone could just as easily call. The spatial anxieties noted in these two examples match the distance currently applied to our asynchronous forms of communication. Texting is a more "absent" way, a more "distant" way, to communicate. Voice conversations are better at giving "presence" because of the synchronicity, the ability to engage in dialog and emulate being face-to-face. However, as noted, each of these notions of absence and presence has deep historic roots.

In my estimation, one of the key concerns that generates these anxieties is the alterations of social space that are produced in conjunction with emerging media. Recently, I was in a local bar with several friends, two of whom were on their phones texting. As they read messages on their phones, they alternated laughter, one after the other. It took me several minutes to realize that they were actually texting each other (even though they were sitting only three feet apart). Proximity here did not accomplish the presence they needed to communicate their ideas. They could have easily have spoken with each other, in whispered tones if discretion was needed. Instead, the space of the mobile interface better served as a private space for conversation. In this scenario, Sherry Turkle's notion of being "alone together" is a bit more nuanced: the mobile phone serves to produce a space that is both private and public simultaneously. We use our mobile technologies to reach out, to connect, and to communicate. At the same time, we are using devices that typically resist group involvement: it is difficult for a crowd to gather around a mobile phone screen to all share in a common experience. However, crowds frequently use their *own* mobile phones (designed, in fact, for individual consumption) to connect in common experiences. Here, the anxiety between presence and absence is mirrored in the anxiety between proximity and distance. The connotations of these words and, in fact, the very practice of these words has troubled their division as distinct and separate ideas.

Locative Mobile Media

To demonstrate how mobile media trouble the distinctions between presence/absence, proximity/distance, and, ultimately, public/private, we can turn to the ways that people are using location-aware applications to interact over mobile devices. On May 1, 2000, President Bill Clinton opened civilian access to accurate GPS signals. Immediately after the "selective availability" to these GPS signals was lifted, people began experimenting with creative uses of their GPS devices (and later, their GPS-enabled mobile phones). From playful uses, such as the GPS treasure hunt game, Geocaching (launched only days after selective availability was removed), to the explosion of popular location-based services that inform users of geolocated information, location-aware mobile devices have forced a reevaluation of the role of proximity in our connections across mobile media. Computing has been characterized by social interactions since at least 2002 or 2003 (with the launch of social networking sites like Friendster and MySpace in those years), and the sites that have drawn the most traffic through most of the early 2000s fit within the category of "social media." It is clear that the ways we have used our computing devices during this era have been centered on social interaction. The same is true of the ways we use our *mobile* computing devices. The distinct change here is a shift in spatial interaction. One of the key features of online social media was the ability to connect globally; mobile media, however, tend to prioritize the local. From applications that map your location and tell you what stores, historical sites, or friends are nearby to applications that allow users to engage and contribute to the meanings and stories of a certain location, uses of smartphones have taken advantage of their media-specificity by connecting people on the go to information about their location.

As mobile phones became computing devices that were able to access the Internet, the device's combination of being used for voice and text communication, location-awareness, and data retrieval from the Internet offered an important change in the way we spatially interact with each other and with cultural objects. For example, in 2010 the Museum of London released a smartphone application called Street Museum. This app took the museum's collection of historic photographs and paintings and geotagged them with location information. The resulting application utilizes "augmented reality" (AR), which superimposes imagery and information onto the physical landscape through your phone. As someone loads the app onto their smartphone, the phone is able to locate their position and show them which locations nearby have a photograph associated with it. Thus, as someone walks down Queen Victoria Street in London, they can hold their phone up and see an image of the Salvation Army Building from 1941 overlaid onto the current street where this building no longer stands. The image the user sees transposed onto the street is of the building collapsing from World War II bombings, "the most severe attack London had sustained throughout the Blitz," as described by the caption in the application.

These kind of location-aware applications that draw from a rich database transform the user's sense of implacement. In other words, by experiencing a simultaneous layering of space and time, the user is offered a deeper sense of context and meaning for the place that defines their sense of self and body. Many of these applications also allow users to contribute to the information in these databases to expand the context of the place and include site-specific stories that may have never been recorded otherwise. These apps thus serve as an important example of another important shift in embodied space that accompanies an emerging mobile media: the clear lines between mediated space and material space are no longer distinct. A user's everyday physical world is now informed by the way their mobile media contextualize and give meaning to a space. In this way, the production of space is achieved both through a person's embodied sense of the space and through how the media extend and inform that situated perspective.

As users of mobile computing technologies find increasingly innovative ways to attach histories, narratives, and stories to locations, they are also changing the definition of what's important about a space. Site-specific and context-aware mobile media projects seek to address how to best attach a wide array of histories and narratives to a location. For example, [murmur] is a project started in Toronto that places large ear-shaped signs on street corners with a phone number that people can call to hear histories about the place at which they are standing, recorded by people from that community. People can also contribute to the history of the location by recording their own story about that place for others to hear. Mobile and locative media thus address Gitelman's concern mentioned at the beginning of this chapter: instead of putting forward one version of history, these mobile media projects use locative, context-aware computing to offer a multiplicity of meanings of a location.

I end on this example to point to an important idea that I hope will frame my historicization of mobile media: within the emergence and the uses of mobile media are woven many histories. There isn't one history of mobile media, there are many. Drawing again from Gitelman's work on historicizing new media: "Their histories must be social and cultural....Any full accounting will require, as William

Uricchio puts it, 'an embrace of multiplicity, complexity and even contradiction if sense is to be made of such' pervasive and dynamic cultural phenomena."[28] Thus, as we continue to write the histories of mobile media, it is vital to undo notions of an "official history," which tends to obscure the deeper cultural moments and processes that inform the creation and use of a particular medium. Such "totalizing narratives," as performance studies theorist David Román suggests, will overdetermine the arrival and meaning of our cultural objects.[29] Ultimately, such totalizing histories will flatten and erase the rich and varied histories that inform and give meaning to our mobile media.

Notes

1. Bianca Bosker, "Foxconn Suicides Continue: Inside The iPhone Factory Plagued By Deaths," *Huffington Post,* May 27, 2010, http://www.huffingtonpost.com/2010/05/27/foxconn-suicides-continue_n_591616.html#s94014
2. Rich Ling and Jonathan Donner, *Mobile Communication* (Cambridge, UK: Polity Press, 2009), 16.
3. Lisa Gitelman, *Always Already New: Media, History, and the Data of Culture* (Cambridge, MA: The MIT Press, 2006), 1.
4. Ibid., 2.
5. N. Katherine Hayles, *Electronic Literature: New Horizons for the Literary* (Notre Dame, IN: University of Notre Dame Press, 2008), 104.
6. Harold Innis, "Media in Ancient Empires," in *Communication in History: Technology, Culture, and Society,* 6th ed., ed. David Crowley and Paul Heyer (Boston: Allyn and Bacon, 2007).
7. For further details about the mobility and the social transformations of the printing press, see James Burke, "A Matter of Fact," *The Day the Universe Changed,* DVD, directed by Richard Reisz (London: BBC, 1986). See also Lewis Mumford, "The Invention of Printing," in *Communication in History: Technology, Culture, and Society,* 6th ed., ed. David Crowley and Paul Heyer (Boston: Allyn & Bacon, 2007).
8. "Remediation" is a term coined by Jay David Bolter and Richard Grusin in their book by the same name (Cambridge, MA: The MIT Press, 1999). Throughout this work, they note that remediation is more than a simple refashioning of one medium into another, but a more complex incorporation of various media into a single medium (and the cyclical effects as old media change from forces imposed by emerging media).
9. Allen W. Palmer, "Negotiation and Resistance in Global Networks: The 1884 International Meridian Conference," *Mass Communication and Society* 5, no. 1 (2009): 9.
10. Marcia Amindon Lusted, "The Day of Two Noons," *Cobblestone* 32, no. 1 (2011): 12.
11. Michael Bull, "Investigating the Culture of Mobile Listening: From Walkman to iPod," in *Consuming Music Together: Social and Collaborative Aspects of Music Consumption Technologies,* ed. Kenton O'Hara and Barry Brown (Dordrecht, The Netherlands: Springer, 2006), 146.
12. Ibid.
13. Mizuko Ito, Daisuke Okabe, and Ken Anderson, "Portable Objects in Three Global Cities: The Personalization of Urban Places," in *The Reconstruction of Space and Time,* ed.

Rich Ling and Scott W. Campbell (New Brunswick, NJ: Transaction Press, 2009), 74.

14. Clive Thompson, "How Twitter Creates a Social Sixth Sense," *Wired* 15, no. 7 (2007), http://www.wired.com/techbiz/media/magazine/15–07/st_thompson

15. John Meurling and Richard Jeans, *The Mobile Phone Book: The Invention of the Mobile Phone Industry* (London: Communications Week International, 1994), 43.

16. Jon Agar, *Constant Touch: A Global History of the Mobile Phone* (Cambridge, UK: Icon Books, 2003), 10.

17. Rich Ling and Jonathan Donner, *Mobile Communication* (Cambridge, UK: Polity Press, 2009), 31.

18. Agar, 19–21.

19. Ibid., 38.

20. Ibid.

21. Gary A. Garrard, quoted in Agar, 43.

22. Howard Rheingold, *Smart Mobs: The Next Social Revolution* (Cambridge, MA: Perseus Books, 2002), xi.

23. Tricia Duryee, "A New First in Mobile: Data Traffic Outstripped Voice Traffic Last Year," *mocoNews.net*, April 1, 2010, http://moconews.net/article/419-a-new-first-in-mobile-data-traffic-outstripped-voice-traffic-last-year/

24. This critique is simply a refashioning of Plato's argument in the *Phaedrus* that art and writing (as documents) signal distance and absence since the document can never engage in dialog and answer for itself. It thus offers a false sense of knowledge and connection. This mode of thinking was brilliantly critiqued by Jacques Derrida in his chapter "Plato's Pharmacy," in *Dissemination*, trans. Barbara Johnson (Chicago: University of Chicago Press, 1981).

25. Kenneth J. Gergen, "The Challenge of Absent Presence," in *Perpetual Contact: Mobile Communication, Private Talk, Public Performance*, ed. James E. Katz and Mark A. Aakhus (Cambridge: Cambridge University Press, 2002), 236.

26. Jason Farman, *Mobile Interface Theory: Embodied Space and Locative Media* (New York: Routledge, 2011).

27. Henry David Thoreau, *Walden: or, Life in the Woods* (Boston: Beacon Press, 1854/1997), 48.

28. Gitelman, 7.

29. David Román, *Acts of Intervention: Performance, Gay Culture, and AIDS* (Bloomington: Indiana University Press, 1998), xviii–xix.

2

Calling Ahead

Cinematic Imaginations of Mobile Media's Critical Affordances

Scott W. Ruston

Amid the annual fanfare of Super Bowl commercials in 2007, one highly anticipated 30-second spot greeted eager viewers and technophiles: The iPhone said, "Hello!" With its montage of 32 clips from film and television with principal characters uttering the typical American telephone greeting, the "Hello" commercial leverages cinematic history to forge a close association of technology and culture. The commercial wraps this association with the wonderment of what this new ringing media device brings.

The association of new technology, especially telephony, and the cultural imagination, rendered through cinema and television, dates to early cinema. Film historians Jan Olsson and Tom Gunning have each addressed the close relationship of the telephone and early 20th century cinema. While Olsson offers a detailed analysis of how early cinema depicted telephone calls, Gunning suggests that the telephone aided new cinema viewers in comprehending the contemporaneous spatial shifts implied by cinematic editing.[1] By showing a telephone conversation in either split screen or parallel editing, early films taught viewers about critical capabilities, techniques and developing conventions of the new medium of film.

Contemporary media users are no less in need of guidance than film audiences in the early days of cinema one hundred years ago. Only today, the roles are reversed. Cinema and television are the known quantities, and mobile media is the new frontier without fully understood or developed capabilities, techniques and conventions. Underneath the utopic zeal that the wireless-communication and consumer-electronics industries use to herald new products and services, there lies a deep anxiety about what services will succeed, which will fail, and ultimately what role mobile devices will really play in our lives. A conceptual framework to distinguish the unique characteristics of this new medium is required.

With the advent of television, a similar need for a conceptual framework existed. "Liveness," as a marker of distinction separating television from film, has long been a cornerstone of television studies. Indicating television's capacity to broadcast live events, as well as capturing an aesthetic of the seemingly live (think of the "live" television studio audience of game shows and sitcoms), early uses of the concept served to explain the uniqueness of television.[2] Even Jane Feuer's landmark 1983 critique of liveness failed to remove the concept from the field; rather, it solidified the term, moving the concept from an idea born in the technology of constantly scanning electronic beams to one of ontology and ideology.[3] Feuer moved the field from a purely formal consideration of the essence of the medium to a technological and cultural formulation. Liveness remains an important critical lens through which scholars examine contemporary television.

As mobile phone market penetration soars, the techno-cultural process of convergence accelerates and entertainment/art options with mobile devices proliferate. How can we understand the essence of mobile media, particularly as it relates to the mobile phone as an entertainment platform? What sets it apart from its entertainment media forebears, especially film, radio, television and the personal computer? In short, the study of mobile media needs an equivalent to television's "liveness."

I argue that mobile media exhibits a unique combination of *affordances* and that we can see mobile media's differences from other communication and entertainment media in those affordances and understand the intersection of mobile media and culture more deeply. Design theorist Donald Norman's notion of *affordance*, popularized in his book *The Psychology of Everyday Things* (later revised and reissued as *The Design of Everyday Things*), expands on psychologist J.J. Gibson's physical phenomenon and "refers to the perceived and actual properties of the thing, primarily those fundamental properties that determine just how the thing could possibly be used."[4] Norman's idea, however, focuses on evident characteristics of technological objects and their use (knobs are for turning, balls are for bouncing, buttons are for pushing, etc.). Much like Feuer's view of liveness, though, I conceptualize affordance as encompassing characteristics beyond the technological and physical, and into the cultural. While obvious once considered, the affordances of mobile media are features that often go overlooked. For example, affordances help us to understand the different contexts by which mobisodes (i.e., unicast, view-on-demand clips of televisual content) and mobile TV (i.e., multi-cast or broadcast televisual content) might satisfy different user desires, and what social practices are linked to each form of mobile entertainment.

There are five critical affordances of mobile media. By "critical," I am referring to the sense of offering tools to critique and analyze, but also to the sense of being important and central to the operation of the media system. These five affordances are:

- Ubiquity—that quality of being always in our presence, always in our possession, and always connected to a larger mediascape.
- Portability—the ability to take a device with us wherever we go facilitates ubiquity, and the combination of portability and connectivity facilitates locativity (see below).
- Personality—an affordance that encapsulates individuality and bodily materiality; the concept extends beyond the mere customization of physical traits, such as bedazzled handsets (few televisions and personal computers exhibit similar levels of decoration) and uses (widgets and apps) to an overall cyborgian quality of an extension of the self and body; we need only consider the term for the mobile handset in Finland (*kännykkä* [can-yuh-**keh**] or "handy" or extending of the hand)[5] or innumerable "sexting" scandals to recognize that there is a close association with the body, intimacy and identity.
- Connectivity—the technological connection of handset to cell tower to switching infrastructure to Internet by way of RF energy, as well as the more important social sense of human interaction and engagement. This latter sense helps explain why many students answer "Facebook" when asked what sorts of mobile media entertainment they engage in.
- Locativity—the capacity of the mobile phone to be out in space, unbound from a fixed location, unlike the television and cinematic experience, and make a relevant association between media content and place for the user. The capacity of a mobile device to be location-aware offers opportunities for new engagements between the mediascape and the landscape, and through the affordance of locativity, the mobile device can unite content and place, as well as form a bridge between the physical and the virtual.[6]

The case of mobile TV exemplifies the need for an analytical and conceptual toolkit, such as I propose by using affordances. While on the surface, porting the dominant entertainment and mass communication form of the 20th century to any screen device that can support it seems like an intuitively successful concept, the reality is far different. Successively, 2007, 2008, 2009 and 2010 were each branded "the year of mobile video," yet subscriber rates in the United States for mobile video services remain less than 10 percent of total mobile phone subscribers.[7] Further suggesting that mobile media is not a fully understood and fully developed ecosystem is the demise of Qualcomm's Media FLO service. This technology powered AT&T's Mobile TV and Verizon's V CAST TV mobile video services (as well as an extremely short-lived service marketed directly by Qualcomm). The service never garnered a significant subscriber base, and it ceased operation on March 27, 2011.[8]

Yet, at the same time, numerous technology wireless industry news sites and blogs report that YouTube receives 200 million views per day on mobile devices,[9] and that 75 percent of mobile YouTube viewers watch content *primarily* with their mobile devices (i.e., prefer mobile over personal computer, gaming console or digital set-top box).[10] Clearly there is a disconnect in our media culture between user practices, user desires, and industry development.

Where to Look for Affordances: The Cultural Imaginary

Affordances exist in the cultural imaginary; examples can be found within social practice and they are invoked in experiments by artists, design labs and the like. This chapter primarily focuses on how cinema, television, and advertisements have imagined these affordances, while acting as repositories and expressions of the cultural imaginary. Tom Gunning has suggested that when cinema was an emerging new media it used the telephone "to naturalize film's power to move through space and time."[11] Similarly, cultural theorist Marsha Kinder contends that the Surrealist films of Buñuel, Rivette and others offer rich insights into conceptualizing narrative for the non-linear and interactive form of electronic media.[12]

While Gunning argues that telephone technology taught audiences about cinema, Kinder suggests that cinema can (perhaps unknowingly) teach us useful lessons about interactive media. In other words, the telephone assists in the naturalizing of cinema's control of space and time, and reciprocally, cinematic representations imagine and naturalize the culturally relevant features of the mobile phone. In short, early cinema viewers could look to their experience with telephones to understand cinema; we can now look to cinema and TV to explain the mobile phone.

By looking at films, television programs, and commercials, we can see the functions and roles that these devices can play and how our popular culture imagines them functioning. This approach shares certain similarities with Kim Sawchuk's study of *Life* magazine advertisements for wireless radio communication in the postwar era. In her study, Sawchuk observes: "What is being sold here is not a machine or a piece of hardware but the affective qualities associated with the technologies."[13] Beyond "what is being sold," my conceptualization of affordance incorporates these affective qualities, along with the specific device capabilities, and, most importantly, helps us understand the entire system of mobile media and their role in our lives.

Two commercials for Verizon's 4G LTE service (providing faster data speeds than the existing 3G system), one for the HTC Thunderbolt handset and one simply for Verizon's service, both harness cultural tropes and affective qualities to suggest specific capabilities and socio-cultural value. The Thunderbolt commercial leverages iconic science fiction imagery, lightning to suggest creation (in *Frankenstein* the doctor stimulates the monster to life with lightning), manipulating time (in *Back to the Future* Marty and Doc capture lightning to power the DeLorean time machine), and opening up other worlds (in *Ghostbusters* lightning crackles around the urban skyline, heralding the arrival of Gozer the Gozerian). The Verizon 4G LTE commercial pairs a farm boy running to the mailbox (reminiscent of *Forrest Gump*) with the conferral of a Zeus-like ability to wield a lightning bolt by the 4G device found in the mailbox. Speed and power, on a mythical and futuristic scale, are the affective qualities conveyed by these commercials. At the same time, they tap into familiar tropes resident in our cultural imagination to express effective qualities of time mastery, access to other worlds, and the power to create. This combination of the affective and the effective lies at the heart of my conceptualization of affordances for mobile media.

Ubiquity

While each of the affordances discussed here shapes the mobile media system and our interaction with it, and plays a significant role in differentiating mobile media from other communication and entertainment media, none is more central than the quality of *ubiquity*. To a greater or lesser degree, each of the other affordances either leverages or becomes more significant by the fact that our mobile media devices, in particular our mobile phones, are nearly always with us and have achieved nearly complete market penetration (in the U.S.). For example, ubiquity and personality go hand-in-hand. While the Finnish colloquial term *kännykkä* emphasizes the personal, even the body, the Japanese language encodes an association of both ubiquity and the personal as *keitei*, variously translated as "handy" or "something you carry with you."[14]

Contemporary popular culture, however, imagines ubiquity as more than the simple constant presence of the mobile phone in one's pocket, belt, or bag. Implied with this presence, and imagined in various ways, is the simultaneous presence of the entirety of the mediascape (which also interconnects the affordance of ubiquity with the affordance of *connectivity* as discussed below). This ubiquitous, omnipresence of the mediascape was the central marketing theme for the sports television network ESPN's short-lived venture into the MVNO business.[15]

Titled "Sports Heaven" and aired during the 2006 Super Bowl, ESPN Mobile's commercial imagines a world with ubiquitous access to sports information. The central figure in the commercial leaves an office building and enters a world where pedestrians are basketball players, base runners, and gymnasts; where city traffic consists of race cars and motocross bikes; and where alleyways are bowling lanes, golfers tee off on sidewalks and bands march through parks. The concluding voiceover proclaims, "Sports fans, your phone has arrived," intimating that a constant connection to the world of sports constitutes a heavenly state of existence for a sports fan, and that the ubiquitous presence of the mobile phone on the owner's person will provide such a blissful state of being. The subsequent failure of ESPN Mobile as a business venture had less to do with a proper leveraging of the affordance of ubiquity, and more with pricing, marketing and handset choice.

Ubiquity cooperates with and facilitates convergence. It is now nearly impossible to purchase a cell phone without a camera function, and this is one of the most successful examples of convergence and leveraging of ubiquity. For many, photo opportunities are missed because an unanticipated event occurs and a camera is not at hand. However, the mobile phone, made ubiquitous because of its portability and connectivity, becomes even more valuable as it ensures that the user is never without a camera for those spur of the moment events. The "Remote Control" commercial for the 2011 Chevrolet Cruze showcases another example of ubiquity and convergence. In this commercial, two men stand in front of a Chevy Cruze, apparently unable to unlock the car doors. One man calls a woman (presumably his spouse) and asks her to "do it again," whereby she activates an "app" on her smartphone, unlocks the doors, and even remotely starts the engine of the car. The commercial suggests that, while someone might forget or lose their car keys (or lock them inside), no one will forget or lose track of their mobile phone. By virtue of its ubiquity, the mobile phone can be the command center for your entire life.

The omnipresence of the mobile phone has caught the attention of scholars, journalists and lawmakers. Many states have enacted laws governing mobile phone usage while driving and most prohibit utilizing non-voice functions, such as texting, while operating an automobile. Scholars such as James Katz and Richard Ling have studied the level of social interruption and the changes to social interaction associated with the mobile phone.[16] In late 2010, Microsoft's "Really" commercial, itself advertising the advent of the Windows Phone, explores the dark side of mobile phone ubiquity. Short vignettes of simple mobile phone usage escalate into scenes of social transgression because of the constant presence of the mobile phone: joggers collide, brides text while proceeding down the aisle, and husbands ignore lingerie-clad wives. Indignant and affronted clients, kids, and businessmen, express their disapproval with the aptly timed and dripping-with-sarcasm interrogative, "Really?" The fact that this commercial offers a critique of what has become a social norm—the omnipresence of and the over-zealous attention to the mobile phone—further cements ubiquity as a well-understood and central feature of the mobile phone.

Portability

The most obvious and overlooked affordance is portability. The small size and light weight of modern mobile phones significantly affects the way we use them, and the role the mobile phone plays in our culture. Of course, the value of portability and ubiquity predates the advent of mobile phones. The opening scene of the first episode of *Get Smart* introduces an iconic symbol, originally of Agent 86's ineptitude and lack of social grace, but now seemingly prescient of society's demand for constant telecommunication access and wearable computing: the shoe phone.[17] This scene, in which Agent 86 receives a call from the Chief while attending an orchestra concert, also forecasts one of the striking social tensions introduced by telephonic ubiquity: the ringing phone in public places. As both the ESPN "Sports Heaven" and the Chevy Cruze "Remote Control" advertisements suggest, the power of the mobile phone extends that of Maxwell Smart's shoe phone: both can travel with you on the sidewalk or onto an airplane, but today's mobile phone also offers access to the world of sports or can unlock a car, or any number of capabilities. Portability, though, does not guarantee ubiquity; rather different uses (and users and designers) of mobile media leverage affordances in different combinations. However, portability is certainly a factor facilitating ubiquity. Without the convenient portability of the mobile device (whether phone, music player, PDA or the like), many of the situations depicted in the "Really?" commercial would not arise. For example, jogging with a large "brick" type mobile phone such as the Motorola DynaTAC 8000, an iconic model requiring nearly two hands to hold, would be nearly impossible, as would simultaneously using the phone and giving a massage.

The DynaTAC 8000 appears in what is purported to be the very first television commercial for mobile phone service (Centel, circa 1989), which imagines associations between portability, leisure, wealth and family. The commercial follows a businessman departing the city en route to the countryside, edited in parallel with scenes of a woman and child enjoying a retreat there. As the man navigates from the world of business (city) to the world of leisure (countryside), he accesses information (directions at a fork in the road), provides updates on his travel (delayed by livestock blocking the road) and announces his arrival to the woman with his mobile phone. Traveling with a phone implies accuracy (no wrong turns) and efficiency (no waiting at the other end). While the scene depicts characters and settings of wealth (e.g., a weekend getaway, nice cars, powerboating on a lake), the commercial suggests that the qualities of the mobile phone illustrated, especially portability and connectivity, accentuate the heavenly and perfect life led by this family.

Well before Centel saw a market for personal mobile phones, science fiction television was imagining the use, appearance, and value of portable mobile communication technologies. The similarity between the *Star Trek* communicator used by Captain Kirk and the modern day flip phone is no accident and the 1960s television show expressed the utility of a small portable communication device.[18] While Maxwell Smart's shoe was a telephone with an infinite wire, having the mobile communicator on his belt allows Captain Kirk to have quick access to the vast resources of his starship, especially the famous transporter. *Star Trek: The Next Generation* (*TNG*) continues the tradition of imagining useful advances in technology. In *TNG*, an insignia badge worn on the Star Fleet uniform replaces the handheld, flip-top communicator. The badge device operates without any requirement to be held in the hand, prefiguring "wearable computing" and more simply the utility of hands-free devices common today (many of which themselves resemble the earpiece of Lieutenant Uhura in the original *Star Trek* series).

Personality

The iconic imagery of *Star Trek* characters and their mobile technologies, exemplified by Uhura's distinctive earpiece or Kirk and his communicator, points to the close association that our culture has assigned to our mobile media devices and our individual persons. The bedazzled decoration of celebutante Paris Hilton's mobile phone is a visible marker, given her frequent appearances in the tabloid media, of the personal attributes we apply to our phones. Unlike the ubiquitous and uniform Model 500 telephone leased by AT&T to residential and business customers for most of the 20th century, mobile phone systems today support hundreds of models with different physical features and capabilities, many with highly customizable modifications. Aftermarket accoutrements also add to the level of personalization.

Our mobile phones are always with us and they have become expressive of our personality via customized ringtones and playlists, individual covers, skins, and other decorations.

These associations occur at the level of character and personality, at the level of the materiality of the body, and even at the level of intimacy. In his *Cell Phone Culture*, Gerard Goggin analyzes the deep interconnection between self, identity and intimacy with regard to sex and SMS text scandals of popular celebrities. While the tales of footballer David Beckham and Australian cricket star Shane Warne connect the mobile phone with the exchange of intimate communications, Goggin's case study of Paris Hilton illuminates how mobile media are even more closely attached to this idea of the personal.[19] The home sex tape at the heart of the Paris Hilton scandal reveals her using her mobile phone during the most intimate and personal of activities, combining an association with the personal and with new levels of ubiquity.

While blinking earpieces, bedazzled cases and interrupted trysts draw a broad connection between mobile media and the personal or intimate, recent trends in the cultural imagination have taken that association and reversed science fiction's generally dystopic view of technologicization of the body. From *The Terminator* to *Star Trek*'s Borg to *Blade Runner*, contemporary science fiction has viewed the integration of robotic technology and the human body with considerable skepticism. Even *RoboCop*'s triumph of the human inside the machine barely tempers a negative portrayal of technology run amok.

However, television advertising campaigns for mobile phones, and Motorola's Droid series phones in particular, offer a twist on the typical sci-fi gloom associated with human-machine integration. In a commercial for the Droid 2 model, featuring a slide-out keyboard, a businessman begins typing on a Droid 2 handset during a meeting as his fingers and arms morph into robotic appendages reminiscent of Arnold Schwarzenegger's *Terminator* cyborg. As this transformation takes place, the voiceover narration extols the data handling and connectivity virtues of the Droid 2 that are "turning you into an instrument of efficiency." Not only does this commercial suggest that Droid 2 users will be as powerful as Terminators, at least in the business world where data mastery leads to career success, it does so by clearly playing on the underlying sense that the mobile phone is already part of our body, that we are already part cyborg and selecting the correct handset will merely enhance our cyborgian power.

The Droid X commercial "Expedition" even more explicitly references popular science fiction culture, and more explicitly represents the mobile phone as part of the cyborg body. Borrowing heavily from the *Halo* franchise of first-person-shooter video games (as well as the *Alien* films, and the *Terminator* series),

"Expedition" follows a futuristic planetary expeditionary squad that enters an underground bunker to find a mysterious object floating in a force field. One member places his bare arm into the object's center void, only to retract it discovering it has transformed into a robotic arm (very similar to the Droid 2 commercial). The Droid X handset materializes out of the palm of the hand (bringing the Finnish term "kännykkä" to mind). Both of these commercials share the stark lighting and desaturated color palette of many dystopic science-fiction futurist films, calling to mind the cinematic imaginary. But, rather than sharing some ambivalence about future technology, these scenes celebrate the union of the machine with the human, and vividly portray an underlying sense that our mobile phones are part of our bodies and our selves—an important component of the affordance of mobile media that I call personality.

In the Korean film *Phone* (Byeong-ki Ahn, 2002) we can see further evidence of the cultural imaginary's association of the phone and the self. But rather than a physical, cyborgian association, *Phone* expresses this personal quality of mobile media by uniting a mobile phone with the soul of a murdered young woman. As the film progresses, rational explanations of technological malfunction are successively replaced by conclusions that the eponymous phone must be possessed by a ghost. The handset in question, originally procured to facilitate an affair, becomes the means of revenge for the murdered and jilted lover, Jin-hee. Jin-hee's ghost, inhabiting the phone, also saves protagonist Ji-won from attack. While the phone serves as a fulcrum between the supernatural and the rational, its role as a central narrative device and as the tool of agency for Jin-hee's ghost cements an association between phone and self, beyond simply the body and extending to the soul. Whether by virtue of customized features, apps and appearances; whether by linguistic or cinematic unions of handset and body; or whether by linking our phones with our identities and our souls, mobile media are best understood, both in their use and in their cultural representation, as the most personal of media devices and systems.

Connectivity

In addition to capturing the powerful connection between our phones and our selves, *Phone* also hints at the mobile phone's ability to forge connections beyond the simple technological connection of wires, RF energy and voices.[20] Numerous films have embraced electronic technologies as a conduit to the supernatural, forging connections across the divide between life and death (e.g., the police lieutenant's phone call from his dead sister's tomb in Luis Buñuel's *The Phantom of Liberty*, 1975) or across time (time-traveling radio communication in Gregory Hoblit's *Frequency*,

2000). And in *Phone,* the white handset formerly owned by Jin-hee becomes a link between the supernatural world her ghost inhabits and the physical world of Ji-won and the other characters. By seeing the phone (or similar communication technology) as the central node bridging these two worlds or time periods, we can expand our notion of what "connectivity" is beyond the literal and technological, to the artistic, to the representative and the human and social. All of these aspects are regularly imagined in cinema and television, and point to the full range of connectivity that we can associate with mobile media.

It is the social, human-to-human connectivity that is invoked by the Centel commercial described above, and even more so by a film such as the thriller *Cellular* (David R. Ellis, 2004). While the movie purports to be a thriller (and meets all the standard conventions of that genre), the driving obstacle to protagonist Ryan's successful rescue of Jessica Martin is maintaining his cell phone connection—on the surface, it is the technological connection needed, but what Jessica Martin really needs (and what Ryan needs in order to get assistance) is a social/human connection. Each needs somebody to believe them. The cell phone is often criticized for creating a privacy bubble around the user, disconnecting him or her from society, and this film emphasizes the need for human-to-human connectivity.

At the same time that *Cellular* hints at social connectivity beneath its action-thriller surface, other examples within the cultural imaginary explore connectivity and the power that it brings. To return to the cultural imaginary for further evidence of an obsession with presence and immediacy and a relationship to connectivity, I turn to the television show *24.* As fans and scholars know, *24*'s obsession with time, marked visually by its on-screen clock, and use of split screens to show simultaneous action, exhibits a double shot of liveness.[21] At the same time, though, the multiple screens that track multiple phone calls and storylines visualize a database of narrative possibilities/alternatives and a networked connection to multiple "feeds"—the viewer is connected to these multiple feeds just as the hero Jack Bauer depends on his connection to the CTU network facilitated by his mobile phone.

An extreme example of the mobile phone's power through connectivity appears in the second season of *24* when Jack Bauer kills Gary Matheson by telephonic proxy. Projecting himself and his pre-emptive strike mentality into the scene by way of his telephonic connection to his daughter, Jack coaches her through the shooting of Matheson and has completely taken control of the situation, using the phone to turn Kim Bauer into his telephonic controlled avatar.[22] Connectivity equates to power for Jack as he eliminates threats and solves conspiracies, moving the story along and saving the world all through the power of his mobile devices.

The combination of the digital clock and split screens affects the audience metaphor appropriate to this show. Television studies have often positioned the audience as a passive, "fly on the wall" observer, or even an extended member of the live studio audience. The audience metaphor for *24* is more akin to a hive of insects all privy to the same knowledge by some sort of neuro-chemical connectivity or to science fiction aliens networked to a central host. *24*, then, expresses a kind of hyper-connectivity, emphasized by the on-screen clock and split screens, and we can see this replicated in our mobile phones. Most mobile phones indicate time of day just like the show does and offer numerous windows, tiles or icons connecting together the range of media options available. The *24* viewer is constantly aware of, if not necessarily focused on, all the active subplots—that is, multiple narrative feeds— just as today's cell phone owner is cognizant of at least the potential connections he or she could make at any one time, not to mention the ongoing text message conversation probably underway and the barrage of Facebook updates cascading down the mobile screen.

Locativity

While the first four affordances are in many ways evident when thought about, even without the extra emphasis of their expressions in the cultural imaginary, locativity is sometimes the hardest to describe. The advent of location-based services has begun to expose the general public to the capacity of mobile media to connect their information experience to their current location. This technology appears in a variety of applications, to include "apps" like Foursquare (location-based social media) or Urbanspoon (location-specific restaurant recommendations) to the wide range of navigation applications available for mobile devices.

Locativity can also be used to refer to the location-awareness many mobile devices now offer, but is currently underdeveloped in terms of entertainment and information media potential. Part of the reason is an under-recognition of how crucial and how central the broad concept of locativity is to the cultural DNA of mobile media. Telephones have a long history of association with a specific place because of their fixed line limitations—consider the large number of thrillers playing upon the location of the hero, victim, and villain, from the de Lorde rescue narratives appropriated by Pathé (*Terrible Angoisse*, 1907) and D.W. Griffith (*The Lonely Villa*, 1908; *The Lonedale Operator*, 1911) up to the current *Scream* franchise (Wes Craven, 1996–2011).[23] Consider, too, the cultural imagination's association of locativity and connectivity. In "The Phony Alibi" episode of *The Adventures of Superman*, Professor Periwinkle invents a telephone that can instantaneously transport a person through the wires to the destination telephone, literalizing the idea of telepresence (used for nefarious ends until Superman saves the day, of course).[24] In a small way, this episode expressed an imagined capability of telephony to go beyond the aural and the mental, and to unite with locativity and bridge two worlds.

Mobile devices extend this imagined capability to a higher degree of relevance between the user, place and information. *The Matrix* (Wachowski brothers, 1999), for example, offers a sophisticated portrayal of telephony and its role in the virtual world humans are trapped within until young hero Neo can set humanity free. Early in the film, Neo has received a mysterious cell phone; activating the phone, he hears the voice of Morpheus. With guidance provided by Morpheus, Neo escapes (temporarily) the pursuit of the Agents. The information provided to Neo is highly relevant to his situation and to his specific location. In this imaginative representation of mobile media locativity, Neo can be guided from cubicle to cubicle to escape his pursuers (note also how the affordances of portability and connectivity cooperate to facilitate locativity). Throughout the film, mobile phones provide location-relevant information to the characters, including directions to escape portals, location of agents, and location of useful objects and assets. While this information does not come to the character via an on-screen "app," it is nevertheless an imagination of the kind of impact mobile media can have on a user's interaction with a space or location.

David Lynch's *Lost Highway* (1997) showcases the mobile phone in a slightly different light. Here it is a touchstone that disrupts the spatial and temporal logic of traditional Hollywood cinema. Standing right in front of the protagonist Fred, the Mystery Man character says, "I'm at your house right now. Call me," and hands his mobile phone to Fred. The Mystery Man's insistence implies that telephonic evidence, used in countless thrillers from early cinema through to the present day, will trump the visual evidence right in front of Fred. When Mystery Man's voice answers Fred's call to his own home, we are left to conclude that either this film's story world does not play by the same rules of our physical world or that the Mystery Man operates beyond the spatial and temporal limitations of the natural world, or both. In any case, the mobile phone is the touchstone that triggers this initial realization.

In addition to *Lost Highway* being one of the first films to use the cell phone as a bridge between rational and supernatural worlds, it also seems to suggest that the cell phone disrupts the location specificity of the landline phone. In so doing, it seeds the notion of the power of mobile media to move beyond connecting two physical spaces, as the landline telephone does, but rather any combination of physical and virtual spaces (a feature exploited in *The Matrix*). This artistic presentation of the cell phone suggests a simultaneous connection of different types of worlds, prefiguring a 21st century lifestyle that brings the world of business into the space of leisure and the world of news media onto the sidewalk and the world of entertainment media into the palm of the hand—a lifestyle and mode of media engagement dependent on locativity and connectivity.

Conclusion

As with most examples of these affordances, there are evident intersections of locativity and connectivity. *Lost Highway*'s imagination of the Mystery Man inhabiting two locations at once expresses both a disruption of spatial expectations and also telepresence, a form of connectivity. The combination of these affordances is now increasingly being explored. *MyTown*, an iPhone app combining real-world locations, social media and a Monopoly-type game, was developed by Booyah Software in 2009 and boasted over 3.3 million players by the fall of 2010.[25] In the spring of 2010, Foursquare teamed with The History Channel to combine location and media content. As these two examples suggest, most of the innovation leveraging locativity is happening in the iPhone and Android app stores, centering around gaming or marketing (the Foursquare-History Channel collaboration was initially developed to promote History Channel's *America: The Story of Us*).

In addition to games and apps, art projects have been consciously leveraging the affordances of mobile media to create innovative and distinctive media experiences. *Yellow Arrow* (Counts Media, 2004–), for example, originated as a spatial annotation project that took advantage of the ubiquity and technological connectivity of the mobile phone in order to forge a human connection to a specific location.[26] In the project, any individual can place a yellow arrow sticker containing a pre-printed code and upload a short text annotation layering thoughts on top of the location identified by the arrow sticker. Subsequent visitors to the space would access the annotation by sending the code to the *Yellow Arrow* servers via SMS text message, and receive back the original annotation. The project incorporates the spontaneity and unexpectedness of street art by essentially assuming that people encountering the yellow stickers will have a mobile phone (ubiquity) with which to access the information; the annotations provide new, location-relevant information to the passerby-participant (locativity), and start to create abstract and mediated human connections between visitors to a space.

The Canadian project *[murmur]* (Shawn Micallef, James Roussel, Gabe Sawhney, 2003–) operates similarly, but depends on voice annotations rather than short texts.[27] Originated in Toronto and now in cities worldwide, the *[murmur]* team partners with civic arts organizations to gather and record stories told by neighborhood residents, making them accessible to passersby in the actual locations discussed by the resident. *[murmur]* installs a recognizable green sign, in the shape of an ear, on a lamppost, telephone pole or other permanent feature in the pertinent location.

A telephone number and code on the sign provide access to pedestrians interested in the space and the stories. The *[murmur]* website does offer a map, suggesting the possibility of dedicated participation time. However, the distance between locations (and this is especially true of the San Jose, California installation) suggests that the bulk of the intended audience is tourists and even local residents who would come across the green ear signs serendipitously, thus further uniting a combination of ubiquity (for access), locativity (media experience linked to space) and connectivity (neighborhood character and history).[28]

Compare these adroit and conscious incorporations of unique features and characteristics of mobile media that affect how we engage with the device, the media content and the media experience. Ubiquity, portability, connectivity and locativity are all central to what makes *Yellow Arrow*, *[murmur]*, *MyTown* and a growing number of similar projects and products unique. By offering the opportunity to add additional commentary to the space, both *Yellow Arrow* and *[murmur]* incorporate the affordance of the personal as well.

Meanwhile, mobile TV has failed to find a user base. The reasons for mobile TV's demise are complicated and involve many factors including pricing, availability, handset compatibility, marketing and business models (among others). However, also part of that failure was a lack of understanding of the differences between mobile media and television as an art, entertainment and communication platform. Mobile TV certainly leverages connectivity by offering access to a section of the mediascape. But does that connectivity enhance the television experience? Only in the cases when the desire to watch television coincides with a lack of access to traditional television or a relevance of televisual information to user's location. This accounts for why news and weather are the most popular forms of mobile video content. Offering only 16 channels on V CAST TV or AT&T Mobile TV, the Media FLO-powered mobile TV services did not significantly engage with the affordance of the personal. Given the close association of mobile media and individuality as well as human connectivity, the growing popularity of Facebook Mobile and video content linked via social media makes sense as leveraging those affordances in a way that mobile TV could not.

The case of mobile TV points to the value of understanding affordances. First, by exploring how the technology has been and continues to be imagined in popular culture gives insight into the broader themes about the roles and associations society has already ascribed to mobile media. And second, understanding those properties that help determine how a technology, in this case mobile media writ large, might be used illuminates how this new media form remediates prior forms and at the same time forges new capabilities and possibilities for integrating and enhancing our lives and the world around us.

Notes

1. See Tom Gunning, "Heard Over the Phone," *Screen* 32, no. 2 (1991): 184–196; Jan Olsson, "Framing Silent Calls," in *Allegories of Communication*, ed. John Fullerton and Jan Olsson (Rome: John Libbey Publishing, 2004), 157–192.

2. See, for example, Herbert Zettl, "The Rare Case of Television Aesthetics," *Journal of the University Film Association* 30, no. 2 (Spring 1978): 3–8.

3. Jane Feuer, "The Concept of Live Television: Ontology as Ideology," in *Regarding Television: Critical Approaches—an anthology*, ed. E. Ann Kaplan (Frederick, MD: University Publications of America, 1983), 12–22.

4. Donald Norman, *The Psychology of Everyday Things* (New York: Basic Books, 1988), 9.

5. See Steve Silberman, "Just Say Nokia," *Wired* 7, no. 9 (September 1999). http://www.wired.com/wired/archive/7.09/

6. See also Scott Ruston, "Blending the Virtual and Physical: Narrative's Mobile Future?," *Refractory: A Journal of Entertainment Media* 9 (July 2006). http://blogs.arts.unimelb.edu.au/refractory/

7. The "year of mobile video" claim was either the explicit theme or an oft-repeated refrain at industry conferences such as Digital Hollywood, NAPTE LA TVfest and CTIA Wireless attended by the author between 2007 and 2010. Subscription figures from Nielsen Mobile available at http://blog.nielsen.com/nielsenwire/online_mobile/americans-watch-more-mobile-video-now-than-ever/

8. Verizon added their "V CAST TV" service (powered by MediaFLO) on March 1, 2007 to an already robust suite of audio and video content services branded as V CAST; AT&T supplemented their "Cellular Video" service with "AT&T Mobile TV" (powered by MediaFLO) on May 4, 2008.

9. See, for example, Stephanie Mosca, "YouTube Mobile Viewers Tops 200M Daily," *TMCnet.com*, January 13, 2011, http://business-video.tmcnet.com/topics/business-video/articles/134511-youtube-mobile-viewers-tops-200m-daily.htm

10. David Sims, "YouTube Video Watching Primarily a Mobile Things, Google Says," TMCnet.com, November 18, 2010, http://www.tmcnet.com/channels/mobile-video/articles/118597-youtube-video-watching-primarily-mobile-thing-google-finds.htm

11. Tom Gunning, "Heard Over the Phone," *Screen* 32, no. 2 (1991): 187.

12. Marsha Kinder, "Hot Spots, Avatars, and Narrative Fields Forever: Buñuel's Legacy for New Digital Media and Interactive Database Narrative," *Film Quarterly* 55, no. 4 (2002): 2–15.

13. Kim Sawchuk, "Radio Hats, Wireless Rats and Flying Families," in *The Wireless Spectrum: The Politics, Practices and Poetics of Mobile Media*, ed. Kim Sawchuk, Barbara Crow and Michael Langford (Toronto: University of Toronto Press, 2010), 34.

14. Mizuko Ito, "Introduction: Personal, Portable, Pedestrian," in *Personal, Portable, Pedestrian: Mobile Phones in Japanese Life*, ed. Ito et al. (Cambridge, MA: MIT Press, 2005), 1.

15. An MVNO or Mobile Virtual Network Operator is a retail mobile phone service provider that leases spectrum and/or airtime from an established network, essentially selling a branded service that piggybacks on a major carrier's network. Boost Mobile, Helio and Amp'd Mobile were well-known "lifestyle" branded MVNOs, while Credo is a "socially conscious" MVNO and ESPN Mobile was a sports-branded MVNO. Helio was acquired by Virgin Mobile (itself an MVNO on the Sprint network); Amp'd and ESPN have both ceased operations.

16. See, for example, James E. Katz, *Magic in the Air: Mobile Communication and the*

Transformation of Social Life (New Brunswick, NJ: Transaction Publishers, 2006); Richard S. Ling, *The Mobile Connection: The Cell Phone's Impact on Society* (San Francisco: Morgan Kauffman, 2004).

17. Mel Brooks and Buck Henry, "Mr Big," *Get Smart,* season 1, episode 1, aired 1965 by NBC-TV.

18. Katz, 68.

19. Gerard Goggin, *Cell Phone Culture: Mobile Technology in Everyday Life* (New York: Routledge, 2006), 126–142.

20. See Jeffrey Sconce, *Haunted Media: Electronic Presence from Telegraphy to Television* (Durham, NC and London: Duke University Press, 2000), for an insightful history of electronic communications and the otherworldly. Sconce suggests that early, point-to-point technologies (telegraph, ham radio, etc.) were imagined as communicating with multiple other worlds and dimensions, while broadcast technologies replaced this vision with the construction of a single, pervasive world. I contend that mobile technologies, combining point-to-point connectivity, broadcast connectivity and locativity, invoke both associations simultaneously.

21. See, for example, the four chapters devoted to style, aesthetics and liveness in *Reading 24* (2007), ed. Steven Peacock (London: IB Tauris).

22. Joel Surnow and Robert Cochran (executive producers), "Day 2: 5:00am-6:00am," *24,* season 2, episode 22, aired 2003 by Fox Television.

23. For a detailed analysis of the role of mobile phones in the horror genre, see Alison Whitney, "Can You Fear Me Now?: Cell Phones and the American Horror Film," in *The Cell Phone Reader,* ed. Anandam P. Kavoori and Noah Arceneaux (New York: Peter Lang Publishing, 2007), 125–138.

24. "The Phony Alibi," *The Adventures of Superman,* season 5, episode 9, aired 1957 by Warner Bros Television.

25. "Booyah's MyTown," Apple iTunes website, accessed May 26, 2011, http://itunes.apple.com/us/app/mytown/id340564769?mt=8

26. See http://global.yellowarrow.net

27. See http://www.murmurtoronto.ca and http://sanjose.murmur.info

28. For a more detailed analysis of these and other location-based mobile media art/entertainment projects, see Scott W. Ruston, "Storyworlds on the Move: Mobile Media and Their Implications for Narrative," *Storyworlds: A Journal of Narrative Studies* 2 (2010): 101–120.

3

Analog Analogue

U.S. Automotive Radio as Mobile Medium

Matthew A. Killmeier

Good fences make good neighbours
—Blum's *Farmer's and Planter's Almanac*, 1850

New and mobile media have pasts that prevailing critical and popular perceptions often scant. Assessing their historical characteristics can provide valuable contexts for gauging their contemporary iterations. In this chapter I assay the history of postwar U.S. automotive radio. This mobile medium—defined as communications that facilitate psychic and accompany physical mobility—provides an analogue for subsequent developments.[1] Although my focus is necessarily narrow, I argue that unifying and segmenting social forces shaped automotive radio. Therefore automotive radio expressed and complemented dialectically intertwined centrifugal and centripetal tendencies. I supplement this specific argument with three more general theses aimed at amplifying historical understanding of mobile media. First, mobile media build upon existing communications linkages, but over time reconfigure them as new communications exchanges. Second, successful mobile content complements the particular characteristics of the mobile medium's development and cultural usage. Third, mobile media are characterized by mobile privatization, mutually constitutive with segmentation and increasing everyday and aggregate mobility.

My approach rests in part on two heterodox premises. First, transportation and communication, or communications, were not sundered with the telegraph.

Jonathan Sterne argues this claim of separation, advanced by James Carey, reduced the holistic conception of communications as transportation and communication to symbolic action.[2] Sterne argues instead for communications as "organized movement and action" that is physical and symbolic.[3] From this perspective we can assess the "greater interconnection, intensity, and reflexivity" that communications foster, particularly so with mobile media.[4] This abets our apprehension of the communicative dimensions of transportation and mobility, and the transportation or mobile dimensions of communication.

My second, related premise counters the ahistorical sensibility that implicates new or mobile media in social change. Media are as much, if not more, affected by the socio-cultural, spatial and economic contexts in which they develop than the converse. I therefore privilege the continuities in social organization, how mobile media are used to reinforce deep-rooted American dispositions.

From its origins, historian Robert Wiebe argues, the U.S. has been a highly segmented and mobile society, and space, capitalism and political organization are chief constituents.[5] The settling and linking of space color American history from westward expansion to 20th century suburbanization. Postal routes, canals, railroads, telegraphy, highways, radio and cars are means of binding space and mobilizing it. They are also means of segmentation. Throughout history "what held Americans together was their ability to live apart."[6] The Pilgrims, utopian communities, company towns and postwar suburbs exemplify how America has consistently "depended upon segmentation."[7] The country's spatial expanse meant "differences were spread across space rather than managed within it."[8] American communications are historically spatial, concerning the physical and symbolic connecting *and* segmenting of people and places.

U.S. capitalism cultivates mobility and segmentation. Its economic imperative has always been to "to grow or die," which spurred spatial expansion and population mobility.[9] The creation of spaces to reduce geographic and temporal fetters on circulation encouraged increased human mobility.[10] As capitalism became mobile so did American society. Industrial capitalism created further segmentation through a division of labor and socio-economic classes. Mass production fostered segments based on consumerism. The U.S. was tailor-made for an economic system dependent upon keeping workers separated from one another but unified by market mechanisms.

American mobility and segmentation were also promoted by its peculiar political federalism. With powers split amongst national, state, county and local governments, unified, central political-economic power was scant. For much of its history the U.S. lacked a metropole to serve as source of national culture, information and economic power and influence. Washington was the center of national politics but little else. When New York began to fulfill this role, it was tempered by "a sturdy tradition of plural centers, an assumption of many norms exemplified in many places, [which] restrained its importance at least as much as its separation from the nation's capital."[11] The plurality of centrifugal metropoles cultivated a nation of regions. As communications facilitated territorial expansion of intercourse, regions became the networks of physical and symbolic mobility through which goods, services, information and people circulated. Metropoles and their satellite regions created unity, but the segmented communities therein limited it. Boston may have been the metropole of New England, but local control, culture and identity exerted strong centrifugal influence over the citizens of Portland, Maine, and other New England locales. "'Born free' in America meant born in pieces."[12] That the pieces could be stitched together as regional agglomerations did not undermine their power. "Segmentation both defined the need for cohesion and set the terms for satisfying it."[13]

Space, capitalism and political organization fostered America's mobility and segmentation. Communications have concomitantly been shaped by these characteristics. They have been a consistent preoccupation as means of unifying the nation and maintaining its segmentation; mobile media are similarly affected.

Communications Linkages to Exchanges

Mobile media build upon existing communications linkages and over time reconfigure them as communications exchanges. Communications linkages are physical and symbolic paths that express and structure circulation and are generally characterized by centrifugal tendencies. They are intertwined with local community and a limited region. Linkages connect particular places. Place "implies an indication of stability" and limitation; that is the spatial character of community.[14] Communications linkages connect but do not greatly affect the particularity of places. Linkages limit the territory, volume and speed of circulation.

Communications exchanges are physical and social nodes that graft onto, overlay and intertwine linkages, connecting them with other linkages and facilitating the expansion, rapidity and mass of circulation. They are generally characterized by centripetal tendencies. Exchanges are mutually constitutive with society. They build upon and connect places and open them up. Space "is composed of the intersection of mobile elements" and exchanges spatialize places, encompassing them within larger circuits of circulation.[15] Exchanges facilitate intercourse within larger geographic units—regional and national. Places are affected in terms of spatial relations and the volume and speed of circulation.

Radio's development illustrates the distinctions and relations between communications linkages and exchanges. Beginning in the late 1920s, radio networks developed as communications exchanges that overlay earlier exchanges. Networks were constituted of stations in particular geographic markets that disseminated much of the same content. Geographic markets built on linkages that were earlier spatialized by communications. National networks were an agglomeration of stations in the biggest markets. Regional networks were constituted of stations in the biggest markets within a large state or region. They comparatively facilitated centrifugal tendencies while national networks facilitated centripetal ones.[16]

In the postwar period radio became more local and regional and less national. This shift had roots in prewar radio. Well-established regional networks and local stations, and non-national advertising co-existed alongside national network radio.[17] Political-economic developments in part contributed to postwar radio's structural changes. First, the radio networks developed television by financing the new medium with radio revenues. National advertising and programming resources increasingly went to television. Second, the total number of radio stations increased with most new stations independent of the networks.[18] Stations affiliated with a network declined from 97 percent in 1947, to 46 percent in 1956 and to 33 percent by 1960.[19] Postwar radio was thus more local and competitive. Postwar radio was also constituted by suburbanization and the growth of automobile transportation.

Postwar suburbanization was a remarkable migration of people from established cities and towns to new spaces—a population spatialization. From 1944 to 1954 suburbs grew ten times as fast as cities, with 9 million people moving to the suburbs from cities.[20] By 1960 more Americans lived in suburbs than cities, and by 1966 one-half of workers and three-fourths of people under 40 were suburbanites.[21] Suburbs were demographically segmented—they were overwhelmingly white because of racist policies and practices, and most residents' income was in the top 40 percentile.[22]

Suburbanization was intertwined with the private car. Suburbs existed in earlier periods, but the growth and character of postwar suburbs was tied to the automobile. Car registrations increased in the 1950s by over 21 million, and by 1955 seven in ten American families owned cars.[23] The 1960 Census noted that 65 percent of workers commuted to work by car.[24] The 1956 Interstate Highway Act—the largest public works project in the 20th century—accelerated suburbanization, automotive dependency and mobility, or automobility.

The physical separations of suburbia were mutually constitutive with automobility. Suburbs were spaces predominantly distanced from services, stores and public facilities. Automobility provided connection and simultaneously stimulated territorial and everyday separation. The postwar suburb was akin to a small town in its separation from cities. Conversely, the suburb was dependent upon and intertwined with the city largely courtesy of the car. The postwar suburban/urban corridor also exhibited some of the frontier/back tier relations of earlier American history. The frontier was subordinate to the back tier, serving as source of wealth extraction for the back tier. It likewise relied upon the back tier for culture and information.[25] The suburban frontier also remained somewhat dependent upon the back tier for culture and information, although not to the extent of the earlier frontier (and less so, in some respects, over time). However, the suburban frontier reversed the extraction of wealth. Residents mostly worked in the back tier but a good deal of their wages and taxes went to the suburbs. Suburbs had local political control over certain facets of their citizens' lives, but regional and national control over culture and consumption was effectively grafted onto and increasingly cannibalized them.

Postwar suburbanites also differed from their predecessors in that they were more mobile every day and in the aggregate.[26] By the late 1950s over 30 million Americans moved each year, with the average American moving every five years and the suburbanite every three.[27] Suburbs were less places than mobile spaces, nodal points where automobility temporarily ceased. Like the suburban home, the car and automotive radio "mediated the twin goals of separation from and integration into the outside world."[28]

Automotive radio effectively complemented the ways Americans were living by facilitating the spatialized regional corridors as communications exchanges, supporting segmentation with a modern twist. For suburbanites the new frontier needed to be navigable; automotive radio served as cultural map. It facilitated and helped symbolically constitute the ties that made the corridor a meaningful interstitial agglomeration, segmented space that retained key divisions yet functioned as a unity. Suburbia's essence was segmentation; therefore common ground was necessarily thin—centering on consumption, lifestyle, news, culture and driving.

Suburbanites desired a distinct form of separation, a division of living intertwined with urban centers and corridors. Automotive radio functioned as part of a communications exchange that knit suburbanites together while helping them stay segmented. People in private cars could maintain their separation from others while connecting to the region in certain ways through automotive radio. It could foster a "community" of listeners whose everyday lives were shaped by spatializing social forces. They stayed in touch without touching.

Form and Content Homologies

Successful radio content complemented the particular characteristics of automotive radio's development and cultural usage. Although automotive radio emerged in the late 1920s, it was a minor facet of radio until after the war.[29] Its development as a unique cultural form lagged in part due to technical limitations. Automotive radio was technically inferior until transistor models began to appear in 1947. Earlier models used vacuum tubes that consumed much electricity, generated a lot of heat and were fragile and bulky. They required a separate battery. High operating temperatures made them unreliable. Prewar automotive radios were also expensive after-market items in most cases, and reception was impeded by the common placement of antennae underneath the car. These stymied the actual and potential automotive radio audience. In the postwar period, total automotive sets in use and the percentage of cars with radios increased significantly.[30] The amount of automotive radios in 1958 exceeded the total radio homes in 1947.[31] A 1949 study of the New York market found that automotive radio usage was substantially greater than usage of home sets during the morning and afternoon—periods later dubbed "drive time."[32] By 1961 nearly one-quarter of the radio audience listened in cars.[33]

Most importantly, however, automotive radio was latent until after the war as it lacked complementary content. Prewar programming was primarily constructed for a stationary, home-based audience. The postwar audience, particularly the increasing segment that listened to automotive radio, required different programming. Automotive radio audiences had distinct listening sensibilities shaped by mobility. While driving or riding in private cars, listeners' primary attention was usually elsewhere (especially the driver's). Mobile content that was background accompaniment better suited their divided attention. Programming that complemented driving also served utilitarian needs. Finally, as most driving was short trips, content that reflected listeners' frequent entering and exiting the soundscape was conducive to everyday automobility.

Recorded music, news and information became radio staples in the postwar period and were homologous with mobile listeners. Music had long been the largest form of radio content, but recorded music was relatively rare in the prewar period. Network bias toward live music and artists' demand for broadcast royalties largely kept it off the airwaves. As the networks left radio for TV, and the radio industry reached agreements with the recording industry, recorded music became economically viable. By 1954 more than half of the average station's content consisted of record shows, and music far outpaced other content.[34] A 1957 station poll found record shows the largest content—comprising 69.1 percent of weekly programming followed by news and weather at 6.1 percent.[35] The length of popular songs, ranging from two to four minutes, suited mobile listening that took place in intermittent chunks rather than longer, linear segments. Popular music likewise did not demand concentrated listening; it was characterized by repetition—a verse-chorus-verse structure—rather than linear narrative. The recording industry explicitly recognized car listeners too, with some labels engineering recordings for the limited fidelity of automotive radio.[36]

Like music, news was a longstanding radio feature. What distinguished it in the postwar soundscape was its integration into broadcast flow, speed of dissemination and attempt to create a strong connection with listeners. Prewar radio news was primarily broadcast as a stand-alone program. Postwar news was often a regular part of each hour's programming. News segments were shorter in length overall—generally five minutes—as was each piece. A 1958 ad for K-JOE (Shreveport) noted, "The old-fashioned newscasts have been replaced by the 'running account' concept," with bulletins sprinkled throughout the hour, "KWICKIE headlines" at half past the hour and "MARK '55'" news five minutes before the hour.[37]

Radio news was also rapidly crafted and disseminated, driven in part by increases in the number of stations after the war and the new technology of magnetic tape. From 1945 to 1956 the number of AM stations increased from 919 to nearly 2,824, and individual stations faced concomitant competition.[38] One way stations stood out was in scooping their competition in news. Furthermore, tape recording appeared after the war and many stations quickly adopted it for news. It helped get news on the air more quickly, giving stations a competitive advantage over television and newspapers. In conjunction with mobile news units, which many postwar stations created, tape recorders were instrumental to the frequent on-the-spot form of radio news. On-the-spot reports bolstered the sense of immediacy and cultivated a desire for mediated communion that preserved segmentation. Such reports promised to "take" the listener "there" by reporting on the scene with mobile units and bringing them the voices and sounds of news.

On-the-spot reportage and mobile units complemented stations' efforts to create a connection with listeners. It was partly sought by creating a unique station identity, such as strong disk jockeys and on-air personalities; promotional events, giveaways, and other gimmicks; station publicity by mobile units; and public service campaigns in conjunction with community organizations.[39] Stations also crafted news to foster a sense of connection and intimacy with audiences. Postwar stations became more locally oriented, particularly with news.[40] In many suburbs radio was one of the only sources of local news, and where there was competition with TV and newspapers radio fought over local turf.[41] "A family's interests change when it moves from one community to another," an industry publication noted, "daily papers in many cases aren't able to respond to or reflect the *new habits* and tastes of this transplanted family" (emphasis mine).[42]

Stations also labored to craft a soundscape that sonically connected the listener. For instance an industry article noted that "more and more the present tense and the first person singular are finding their way into newscasts bringing with them the value of projecting the listener into the atmosphere of the event being reported."[43] News was constructed as happening now, and presenters and audiences were addressed as individuals in newscasts. It could sound more local and intimate when stations taped listener "tipsters" and used their voices in the news.[44] More common was "involving the actual voices of the people making the news."[45]

In addition to music and news, postwar radio regularly featured functional information for mobile listeners, such as weather and traffic reports. If these reports indicated radio's shift from a prewar medium of entertainment to a postwar "nervous information system," then the organisms served were the new suburban communications exchanges.[46] Like news, traffic reports were valued for immediacy as well as ephemeral needs. They had fleeting importance for listeners rushing across the expressway. Traffic reports were a regular part of drive time, postwar radio's new prime time comprising morning and evening commuting times.[47] For example, WWJ (Detroit) did traffic reports every 10 minutes during drive time, WKY (Oklahoma City) featured them every 15 and KGO (San Francisco) did a minimum of 19 daily.[48] In 1957 96 percent of Mutual Broadcasting System's affiliates regularly ran road, traffic and weather reports.[49] Akin to the mobile news units, many stations began using helicopters for traffic reports. They were spectacular forms of station publicity and quite functional. As with many suburban matters, the totality of traffic exceeded the segmented apperception of suburbanites.

Automotive radio was mutually constitutive with segmented suburban life. Home and work were territorially separated, which fostered a distinct sense of spaces rather than place. This also challenged, to some extent, one's identity. In the 20th century place waned in importance to identity and work waxed.[50] While the car connected work and home, city and suburb, it was interstitial. It was private, but operated in public. Drivers were separated yet connected, interacting with others in particular ways. Automobility was a means of fostering a subjective sense of freedom that seemed to meliorate the rationalizing pressures of postwar life.[51] However, "the automobile's signification of centrifugal energy was precisely what enabled it to function as such an effective tool for the centripetal regulation" of drivers.[52]

Automotive radio served to connect drivers to suburban and urban spaces, helping them navigate physically and socially. Like traffic helicopters, automotive radio attempted to provide a sonic, panoptic perspective on segmented suburban space, to bring the listener "there" without leaving "here." It aided in unifying suburban segmentation. Automotive radio addressed audiences who were neither home nor at work, but fluidly in between—an interstitially private and public listener. It abetted the physical mobility of travel, making driving perhaps more bearable, and also fostered psychic mobility. The listener may have been driving a banal highway, or stuck in traffic, but entertainment and information could perhaps take them elsewhere.

Automotive radio connected listeners in ways that complemented their separations. In this respect, automotive radio has many historical analogues: community newspapers, black and foreign language newspapers and afternoon editions of newspapers targeting rail commuters. On the other hand, automotive radio's vividness—it brought actual sounds into the car—and its attempts to speak to the listener directly, represented a new derivation. Here automotive radio had a good deal of commonality with postwar TV and radio's "para-social interaction"—the attempt to vivify media content in order to create a sense of togetherness.[53] Automotive radio may have been used (and arguably still is) to construct para-social community, tapping "the native power of radio to involve people in one another" for the particular American ends of "accompanied solitude."[54]

Mobile Medium/Mobile Home

Mobile media develop as responses to change and mobility as lived experiences, adaptations to larger socio-cultural changes, and are mutually constitutive with segmentation and increasing everyday and aggregate mobility. These adaptations are characterized by mobile privatization. Raymond Williams developed mobile privatization to crystallize broadcasting's socio-historical context. It is characterized by "two apparently paradoxical yet deeply connected tendencies of modern urban industrial living: on the one hand mobility, on the other hand the more apparently self-sufficient family home."[55] In contrast to previous public technologies, mobile privatization "served an at once mobile home-centered way of living," and broadcasting "was a social product of this distinctive tendency."[56]

Mobile privatization's focus on the home, mobility and communications is suggestive about the development of automotive radio and the postwar radio communications exchange. The prewar radio exchange was predicated on stationary, home-centered communications with concomitant longer, linear programming. However, radio fostered psychic mobility—listeners could imaginatively travel from their homes; it spatialized the home. Additionally, as prewar radio was largely national it increased listeners' horizontal mobility. More ideas and things—through national programming and advertising—were circulated. Therefore, the postwar radio exchange grafted onto a psychically, horizontally mobile listener and spatialized home and extended them with automotive radio. Its distinction is that it constitutively developed with suburbanization and automobility.

Automotive radio's programming complemented mobility. It suited the sensibilities of physically mobile listeners, and it offered psychic mobility in the car. On the one hand automotive radio publicized the space of the private car, and on the other helped perceptually privatize social, public spaces. For segmented drivers automotive radio kept them plugged in to public discourse. It provided a sonic tether that linked atomized drivers, "connectivity" they would otherwise lack. Driving with automotive radio is as different from driving without it as the talkies are from the silents. Travel becomes potentially doubled, physically and psychically, with each reflexively affecting the other. Automotive radio colors physical travel with psychic mobility. Driving may also limit the psychic travel of radio because it demands a degree of distracted listening, which is reflected in postwar content.

Automotive radio affected the experience and perception of public space by individualizing and privatizing it. As part of the private car, automotive radio made roads "an agglomeration of privately inhabited spaces, mediated by technologies of transportation and communication."[57] As sound is closely associated with our sense of space, automotive radio was a means of aurally asserting some control over space and one's experience of it.[58] It reinforced the sense of individual freedom Americans associated with automobility, perhaps tempering their awareness of how structured driving actually was. One could choose what roads to take in one's own car and what station or stations to listen to. Automotive radio can "create both the experience of being 'cocooned' by separating the user from the world beyond, and a different 'space' whereby the user" partly inhabits the mediated, psychic space fostered by radio.[59] Automotive radio thus reciprocally affected perception of roads and public space. Roads were spaces for the delicate communicative interaction of driving and para-social interaction. Drivers were regularly interpolated as "you" by newscasts, DJ talk, traffic reports and popular music.

The segmented space of the car and automotive radio enhanced a private, individualized sense of roads and public space. They helped make the road/driver interface a potential collage, a tableau for sonic psychic, aesthetic juxtapositions. A road linked Hartford and Boston, but with automotive radio it was a potential mash-up of Pat Boone, Rosemary Clooney and news of a nor'easter. Like non-diegetic film music, automotive radio's soundscape offered affective, reflexive intensification of driving and public space. In helping make the experience of driving and perception of space more private and individualized, automotive radio contributed to a more monadic and nomadic society.

Williams' insights about broadcasting likewise illuminate programming. Rather than discrete content categories, the planned flow of broadcasting is its essence.[60] Flow is driven by commercial imperatives. It aims to gather and retain listeners by ensuring a unifying continuity for seemingly discrete program units and commercials. Indeed, what Williams recognized had been a longstanding radio practice. "'Flow' is also the important element in the 'balanced programming'...[which] is based upon the concept of radio as a personal medium, a constant companion, designed to appeal to local tastes within an individual market...[and] to provide the ingredients that will keep the most people listening all of the time."[61]

Instead of distinctions between music, news, information and commercials, there was a flow between and among them in mobile content. Recorded music was a de facto advertisement for the artist and recording, but also served the station's commercial needs. Its relatively inexpensive cost, breadth of styles and genres and catalogue depth made it a viable means for attracting and retaining broad and narrow audiences. Recorded music helped stations craft identity through sound—as a whole or through targeted genre shows. Stations could capitalize on particular segmented audiences with musical genres because postwar recording shaped mass audiences into particular demographic segments. Targeted genres gathered demographically homogeneous audiences that were interpolated with complementary advertisements. Likewise news and information and traffic reports served to attract and hold audiences, publicize the station and promote sponsors and their products.[62]

Flow is not limited to commercial connections between programming, audiences and sponsors, but is intertwined with the larger social context. Music, news and traffic reports complemented automobility and suburban living, reflecting centrifugal and centripetal tendencies. Recorded music was locally distributed, but most was nationally produced and promoted. So too with the station formats increasingly adopted across the nation in the postwar period, such as music and news and Top 40.[63] Such formats diminished station diversity and the autonomy of programmers and disk jockeys. On the other hand newer, specialized formats, such as country and western and rhythm and blues, were somewhat centrifugal, catering to regional tastes and minority segments.[64] News was likewise centrifugal and centripetal. Local news was a staple and many stations invested in reporting. However, national news was a part of the mix, and stations largely used monopoly wire services. Traffic reports were inherently centrifugal as they addressed a local/regional phenomenon. However, they were centripetal in that traffic was increasingly a national, everyday problem, albeit one experienced locally. Conversely, the adoption of similar means (helicopters) and forms (drive time reports) was a national phenomenon.

Broadcasting developed in large part as a response to social pressures and needs in capitalist societies where increased change and mobility are lived experiences.[65] Change and mobility were quotidian features of postwar experience that affected postwar and automotive radio. Spatialized suburban living and concomitant automobility drove the need for reconnoitering and navigating a new lifestyle and a somewhat new frontier. Suburbanites were less situated in their developing suburban communities because of the mobile, interstitial character of everyday life. They were split primarily between work and home, but also to some extent between culture, recreation and consumption, and home. Their allegiance to and identification with the suburb were weaker as they were spatially and affectively spread across suburban-urban corridors. Increasingly "what one did superseded where one did it as the initial, controlling question," which strengthened horizontal identities and centripetal tendencies.[66] What suburbanites needed was the sense of interconnection and bonds of traditional community free of the obligations of direct, reciprocal intercourse it rested on. That is, a communicative connection consonant with how they lived. Postwar radio developed in part as a homologous response to these pressures and needs, and automotive radio was its ideal type. It unified by providing a sense of togetherness and belongingness that was defined and limited by the segmentation at the core of suburbia.

Suburbanites' needs for unified segmentation were complemented by political-economic needs to gather and hold audiences and markets. Postwar broadcasting accommodated both the needs of suburbanites and business. However, the latter was more determinative than the former as the economic structure of broadcasting, with scant exceptions, remained commercial. The socio-cultural needs of listeners were addressed, but largely as means to commercial ends—attracting valuable segmented audiences to be sold to advertisers. Mobile content suggests that light ties and thin common ground provided the best means of accommodating the needs of listeners within the designs of industry and advertisers. Content that fostered a para-social community served the desire for togetherness and segmentation, and provided a firm basis for attracting audiences and making markets.

As an historical analogue of mobile media, U.S. automotive radio suggests three intertwined general propositions. First, mobile media abet reconfiguring space by cultivating it as communications exchanges. Mobile media build upon existing communications linkages and aid in their reconfiguration. They take root in established environments, and are interconnected with existing physical and social communications. Therefore, mobile media reflexively intensify their communications environments. Understanding mobile media thus entails assaying the communications environment in which they develop, and accounting for how they contribute to making linkages into exchanges. With automotive radio the environment was the suburban-urban corridor, linked by automobiles, roads and the commuting lifestyle of suburbanites. It was interconnected with forms of mobility, mobile living and segmentation. The resulting suburban communications exchange was a para-social community, a form of "accompanied solitude" composed of light ties and constituting a segmented market. In order to comprehend the greater interconnection, intensity and reflexivity that mobile media foster, it is necessary to situate them in their communications environments.

Second, there is a homologous, mutually constitutive relationship between mobile media form and content. In order to comprehend form and content we have to understand the medium as a cultural form. While cultural form encompasses a particular medium with technological features, it extends beyond the medium and its political-economic contextual framework to privilege social practices and signi-fication.[67] In the case of mobile media, cultural form also includes the interconnec-tions of its communications environment. Automotive radio developed as cultural form intertwined with radio broadcasting, which was undergoing political economic change that oriented it locally. Furthermore, it was intertwined with suburbaniza-tion, automobility and concomitant short travel. Automotive radio's content was mutually constitutive with its political economy and social practices. Therefore, apprehending a mobile medium as a cultural form is to perceive it as a gestalt com-prising the medium in its communications environment and its content.

Third, mobile media develop as responses to change and mobility as lived experiences, as adaptation to larger socio-cultural changes. These changes are deeply rooted in the American experience. As a response, mobile media can be a means of psychic and affective connection that meliorate the social effects of change. Mobile media can help maintain ties that change otherwise inhibits. Simultaneously, mobile media abet such changes. They support more fluid, mobile lifestyles. Finally, mobile media are also used to complement longstanding segmen-tation in American society. Amidst the flux of modernity there is strong continu-ity in separation. Therefore, mobile media develop as a response to larger social forces, and are used to mitigate the effects of change through connecting people and maintaining their segmentation. Mobile media may be bridges to connect or fences to separate. Historical perspective is critical to determining which abides.

Notes

1. Matthew A. Killmeier, "Space and the Speed of Sound: Mobile Media, 1950s Broadcasting and Suburbia," in *Transmitting the Past: Historical and Cultural Perspectives on Broadcasting*, ed. J. Emmett Winn and Susan L. Brinson (Tusacaloosa: University of Alabama Press, 2005), 167.

2. Jonathan Sterne, "Transportation and Communication: Together as You've Always Wanted Them," in *Thinking with James Carey: Essays on Communications, Transportation, History*, ed. Jeremy Packer and Craig Robertson (New York: Peter Lang, 2006), 117–135; James W. Carey, "Technology and Ideology: The Case of the Telegraph," in *Communication as Culture: Essays on Media and Society* (New York: Routledge, 2009), 155–177.

3. Sterne, "Transportation and Communication," 118.

4. Ibid., 129.

5. Robert H. Wiebe, *The Segmented Society: An Introduction to the Meaning of America* (New York: Oxford University Press, 1975).

6. Ibid., 46.

7. Ibid.

8. Ibid., 29.

9. Ibid., 22–23.

10. David Harvey, *The Condition of Postmodernity: An Enquiry into the Origins of Cultural*

Change (Malden, MA: Blackwell, 1990), 232.

11. Wiebe, *The Segmented Society,* 37.

12. Ibid., 35.

13. Ibid., 91.

14. Michel de Certeau, *The Practice of Everyday Life,* trans. Steven Rendall (Berkeley: University of California Press, 1984), 117.

15. Ibid.

16. An important exception was the national Mutual network, which encompassed stations affiliated with other national networks, as well as entire regional networks. Jim Cox, *American Radio Networks: A History* (Jefferson, NC: McFarland & Co., 2009), 72–88.

17. Alexander Russo, *Points on the Dial: Golden Age Radio Beyond the Networks* (Durham, NC: Duke University Press, 2010).

18. Susan J. Douglas, *Listening In: Radio and the American Imagination from Amos 'n' Andy and Edward R. Murrow to Wolfman Jack and Howard Stern* (New York: Times Books, 1999), 224.

19. Christopher H. Sterling and John M. Kittross, *Stay Tuned: A History of American Broadcasting* (Mahwah, NJ: Erlbaum, 2002), 830–831.

20. Kenneth T. Jackson, *The Crabgrass Frontier: The Suburbanization of the United States* (Oxford: Oxford University Press, 1985), 238.

21. Ibid.

22. Roland Marchand, "Visions of Classlessness, Quests for Dominion: American Popular Culture 1945–1960," in *Reshaping America: Society and Institutions 1945–1960,* ed. Robert H. Brenner and Gary W. Reichard (Columbus: Ohio State University Press, 1982), 168–169.

23. Douglas T. Miller and Marion Nowak, *The 1950s: The Way We Really Were* (Garden City, NY: Doubleday, 1977), 139.

24. Ibid.

25. Carey, "Space, Time, and Communications: A Tribute to Harold Innis," in *Communication as Culture,* 115–124.

26. Pitirim A. Sorokin, *Social and Cultural Mobility* (Glencoe, IL: Free Press, 1959), 382.

27. Miller and Nowak, *The 1950s,* 136.

28. Lynn Spigel, *Welcome to the Dreamhouse: Popular Media and Postwar Suburbs* (Durham, NC: Duke University Press, 2001), 32.

29. For a sketch of prewar automotive radio, see Russo, *Points on the Dial,* 170–175.

30. *Broadcasting Yearbook—Marketbook Issue* (Washington, DC: Broadcasting Publications, 1960), F-63; Sterling and Kittross, *Stay Tuned,* 862; Radio Advertising Bureau, "Listeners on Wheels," 1953, Box 2, Folder 1, Mass Communications History Collections, Wisconsin Historical Society, Madison, WI.

31. "81 Per Cent More Radio Sets in Use Than in 1948," *Radio-Active,* May 1958, 4.

32. WOR Research Department, "The Early Morning and Early Evening 'Audience on Wheels' in Metropolitan New York, 1949," WOR Station File, Pamphlets (1949–1987), Library of American Broadcasting, University of Maryland, College Park.

33. Sherman P. Lawton, *The Modern Broadcaster: The Station Book* (New York: Harper & Bros., 1961), 42.

34. Philip Eberly, *Music in the Air: America's Changing Tastes in Popular Music, 1920–1980* (New York: Hastings House, 1982), 174.

35. "1957 Disk Jockey Poll," *Billboard,* November 11, 1957, 58.

36. Brian Ward, *Just My Soul Responding: Rhythm and Blues, Black Consciousness and Race Relations* (Berkeley: University of California Press, 1998), 283.

37. *U.S. Radio,* February 1958, 2.

38. Sterling and Kittross, *Stay Tuned,* 827.

39. Matthew A. Killmeier, "Voices Between the Tracks: Disk Jockeys, Radio and Popular Music, 1955–60," *The Journal of Communication Inquiry* 25, no. 4 (October 2001): 353–374; "Local Promotion: Civics or Gimmicks?," *U.S. Radio,* October 1957, 43–45; "Premiums and Prizes in Radio," *U.S. Radio,* May 1959, 53–54; "Mobile News Units Attract Advertiser Support," *U.S. Radio,* November 1958, 53.

40. "Local Radio Phenomenon," *U.S. Radio,* May 1959, 21–24.

41. "Suburbia: Newspapers Miss a Beat," *U.S. Radio,* March 1959, 93–94.

42. Ibid., 94.

43. "Selling the 'Sound,'" *U.S. Radio,* December 1958, 37.

44. Seymour Freedgood, "The Money-Makers of 'New Radio,'" *Fortune,* February 1958, 122–124, 222–226.

45. "Radio Formats: What Is Radio Today?," *U.S. Radio,* May 1958, 20–25, 85.

46. Marshall McLuhan, *Understanding Media: The Extensions of Man* (New York: Signet, 1964), 260.

47. Gerald Bartell, a station owner credited with innovating the successful music and news format many postwar stations emulated, coined the term "drive time" in 1957. Peter Fornatale and Joshua E. Mills, *Radio in the Television Age* (Woodstock, NY: The Overlook Press, 1980), 20.

48. WWJ ad, *Radio-Active,* February 1956, 9; "Public Service—Radio Broadcasters Do Their Part," *Radio-Active,* June 1957, 7; KGO ad, *U.S. Radio,* June 1958, 40.

49. Mutual ad, *Radio-Active,* February 1957, 8. Unlike NBC, CBS and ABC, Mutual never ventured into TV and remained a radio network until 1999.

50. Wiebe, *The Segmented Society,* 24.

51. Cottten Seiler, *Republic of Drivers: A Cultural History of Automobility in America* (Chicago: University of Chicago Press, 2008), 41.

52. Ibid., 143.

53. Donald Horton and Richard Wohl, "Mass Communication and Para-social Interaction: Observations on Intimacy at a Distance," *Psychiatry* 19 (1956): 215–229.

54. McLuhan, *Understanding Media,* 261; Michael Bull, "Sound Connections: An Aural Epistemology of Proximity and Distance in Urban Culture," *Environment and Planning D: Society and Space* 22, no. 1 (2004): 105.

55. Raymond Williams, *Television: Technology and Cultural Form* (New York: Schocken, 1974), 26.

56. Ibid.

57. Seiler, *Republic of Drivers,* 146.

58. In this respect automotive radio anticipated the Walkman and MP3 player. Bull, "Sound Connections," 111.

59. Ibid., 113.

60. Williams, *Television,* 86.

61. "Selling the 'Sound,'" *U.S. Radio,* December 1958, 37.

62. "Mobile News Units Attract Advertiser Support," *U.S. Radio,* November 1958, 53.

63. Freedgood, "The Money-Makers"; Ben Fong-Torres, *The Hits Just Keep on Coming: The History of Top 40 Radio* (San Francisco: Miller Freeman, 1998), 37–46.

64. "How Radio and Country Music Sell," *U.S. Radio,* April 1958, 26–29; "Negro Radio Zeroes in on National Advertising," *U.S. Radio,* December 1959, 24–30.

65. Williams, *Television,* 23.

66. Wiebe, *The Segmented Society,* 24.

67. Raymond Williams, *Culture* (London: Fontana, 1981), 148–180.

4

CB Radio

Mobile Social Networking in the 1970s

Noah Arceneaux

If you pay a visit to the stacks of any reasonably stocked university library (particularly those shelves whose call numbers begin with HE, PN, or TK), you will find a myriad of works devoted to many phases of radio history. One particular topic, though, has received scant attention from media historians despite its astonishing popularity in the 1970s—the phenomenon of citizens band (CB) radio. Along with other fads from that era, such as pet rocks, leisure suits, and disco music, CB has been relegated to the nostalgia bin of history, with one academic study observing that there is "virtually no scholarly literature on the topic."[1] While this characterization is something of an exaggeration, academic studies of this particular medium are relatively scarce in comparison to other aspects of radio.[2]

Scholars' neglect of CB is consistent with Light's observation that media historians have focused the bulk of their attention on technologies that conform to the broadcast, or one-to-many, model of communication, rather than ones that serve inter-personal communicative functions.[3] With the addition of multi-media capabilities to portable phones in the 21st century, however, the presumed distinction between mass and inter-personal communication has blurred. The concomitant rise of user-generated online content, colloquially dubbed Web 2.0, as well as the rapid growth of online social networks further confuse any such distinction. In an overview of the current status of international mobile communications, Goggin identifies the growing phenomenon of MoSoSo, or "mobile social software," whose growth is fueled by the "confluence of two important shifts—from desktop to mobile computing, and from individual to social software."[4] These mobile services allow users with similar interests in physical proximity to exchange information, and engage with such location-based services as geotagging, and will no doubt grow in popularity as the two overlapping trends identified by Goggin continue to flourish.

The dramatic and continuing changes to media technologies, exemplified by the MoSoSo phenomenon, inspire a re-examination of CB radio for at least two reasons. This amateur radio service first gained popularity in the United States within the community of long-haul truck drivers before average citizens began to install the devices in their own suburban vehicles. The link between mobility and CB radio was firmly established in the public consciousness (though nothing about the technology was inherently or necessarily mobile). CB was also situated at the nexus of mass and inter-personal communication; users could broadcast warnings of impending traffic hazards or police speed-traps to anyone within range, while they could also carry on conversations with specific individuals. Neither of these two aspects of CB, its link with mobility or its blurring of mass and inter-personal communication, attracted significant scholarly interest in the past, and this present excursion into media archeology is another indication that history is not a fixed entity as much as a confluence of competing narratives, continually rewritten based on current concerns.

While acknowledging the scarcity of previous intellectual interrogations of CB, this study is admittedly only one step towards addressing the oversight. The phenomenon of CB could be analyzed in relation to a number of topics, beyond those raised within this chapter. There are, for example, the contradictory spatial dynamics at play within the medium. Since the over-the-air messages were typically broadcast for several miles, the physical location of a specific user lost some of its original significance; the fact that most messages dealt with such quotidian issues as traffic conditions or the location of police vehicles, however, magnified the significance of one's location. There is also the noted tension between surveillance and security, a tension that has been linked to mobile phones as well.[5] Such devices can be used to summon safety in times of crisis, or to warn others of law enforcement, though the very act of participating in the network simultaneously makes surveillance of the user also possible.[6] Another parallel between CB and contemporary media technologies is the skeptical attitude from some academics toward the particular social bonds made possible by the airwaves.[7] Such critiques foreshadowed popular culture's frequent denouncement of the nature of online relationships.

While these issues are fertile terrain for exploration, the focus of this chapter is instead on the popular culture hysteria associated with this form of communication in the 1970s. Drawing from previous scholarly literature on the topic, mainstream press coverage, and songs and films that sought to capitalize on the fad, this chapter uses the public response to CB as a historical framework for evaluating contemporary perceptions of mobile phones, social media and online communication in general. Specific themes within the public discourse are identified, with a focus less on how CB was utilized and more on how the technology was imagined, conceptualized, and glorified in public discourse.

Innovations and Exploitation

Before identifying specific themes associated with the fad, it is first necessary to present an overview of CB radio history, a history that shares much with other media technologies.[8] Recalling de Certeau's theory of *la perruque,* the development of this medium was marked by a series of negotiations between users and regulators; adopters failed to use the technology for the purposes it was originally intended, while the government continually attempted to corral an unruly beast.[9] The U.S. military, whose control of radio during World War I led directly to the subsequent growth of broadcasting, was again instrumental. Innovations during World War II opened up new portions of the airwave spectrum, and with the end of hostilities in sight, the Federal Communications Commission (FCC) authorized a Citizens Radio Communication Service using ultra-high frequencies.

The name given to this initial service was something of a misnomer, however, as the new channels were specifically *not* intended to become a national party line or an open-air chat room. The proposal instead was for low-power radios that could transmit for a few miles, and would be used by particular institutions. Department stores might use them to coordinate delivery vehicles, or hospitals to contact off-duty physicians. This original service proved to be far less popular than the FCC had planned, as the only equipment that could operate on these frequencies was prohibitively expensive.

In 1958, the government opened 23 new channels in a lower frequency portion of the spectrum; two-way radios that worked with this service were now affordable. A new category of radio license was created, and obtaining one was simply a matter of paperwork—no money or examination required. Use of profane language was prohibited, and the FCC tried to limit inter-station calling. As originally planned, a "station" might comprise a construction company or a shipping company, and communication was to be restricted to those users who worked for the same station. The service's popularity and the volume of inter-station traffic (strangers talking to strangers) surprised the government, and many users never bothered to obtain the requisite license. The FCC was eventually forced to permit inter-station calling on seven of the channels.

With so many users trying to communicate on a limited number of channels, interference became a problem. A twenty-dollar license fee introduced in 1964 failed to thwart the popularity of the service. Establishing the actual number of CB radio users from this era is impossible, as the license fee was even more encouragement to ignore this bureaucratic obligation. Most users likewise ignored the requirement to identify themselves with specific call letters, and instead adopted colorful nicknames, known as "handles." As there was no official directory for correlating handles with individual users, identifying unlicensed CB-ers was difficult if not impossible. For reasons that are unclear, most CB users did however obey the FCC's prohibition against vulgar language, with the all-purpose word "mercy" substituting for other curse words.

Along with the handles, a distinct jargon evolved for communications via CB, with a white, Southern, heterosexual male imagined as the ideal user. "Good Buddy" became the most common greeting, while women were referred to as "beavers" or "seat covers" and roadside diners denigrated as "choke-and-puke." The vocabulary also incorporated the police ten-codes, which were originally a method for minimizing on-air traffic with common phrases denoted by specific numbers. While the ten-codes retained this instrumental function on CB channels, the arcane vocabulary also delineated regular users from novices. Truck drivers, in particular, were fond of the device as it was a marked improvement from the older system of using hand signals to notify others of road conditions or nearby police.[10] Channel 19 became the unofficial trucker channel, and drivers lamented the influx of average citizens to their turf. Inaccurate use of CB lingo was one way to distinguish the long-time users from the party crashers, though one trucker (identified simply as "Ribbon Wrangler") also claimed the particular vibrations that reverberated through a truck produced a distinctive sound:

> I can tell one of those four-wheeler rubberbanders before he opens his mouth by the key of his mike, the vibration of his engine and let me tell you I don't come back....Tell those four-wheeling "cowboys" to keep off Channel 19![11]

In late 1973 and early 1974, independent truckers orchestrated a number of protests against rising gasoline prices, using their vehicles to block major highways. Some of these protests turned violent, and the use of CB radio became a recurring trope within mainstream press coverage.[12] The connection between this technology and rebellious, anti-authoritarian truckers was cemented in popular culture. In 1974, Congress also lowered the speed limit to 55 miles per hour in order to conserve gas, and truckers saw CB as the ideal method to evade police speed traps.

Overdrive, one of the primary trade journals for truckers, began to heavily advertise CB radios in April 1974, signaling a rapid growth period for the technology.[13] In the ensuing two years, the number of CB radios imported to the United States (primarily from Japan) mushroomed from less than four million to more than ten million, with the phenomenon peaking in 1976–77.[14] As previously noted, establishing the number of actual users was impossible, given the proliferation of unlicensed users, with estimates ranging from 30 to 50 million.[15] Such numbers alone cannot provide the full picture of the medium's popularity, however, as citizens from every segment of society were fascinated by this curious new form of communication. Betty Ford, the First Lady of the United States, even joined the fray, adopting the handle "First Mama" for her transmissions, while drivers who had no desire to actually chat over the airwaves could purchase fake antennas "to give [their] car, truck or van the popular CB look."[16] There was also no shortage of guidebooks and jargon dictionaries to take advantage of the market, and at least two CB-based board games hit the shelves. The television programs *Movin' On* (1974–78) and *BJ and the Bear* (1979–81), both of which starred CB-using truck drivers, bookended the fad, with an animated 1977 cartoon, *CB Bears,* also popping up on Saturday morning TV.

Hollywood studios, true to form, released a number of films to capitalize on the fad with most all of them depicting CB an integral part of the trucking lifestyle. In the drive-in exploitation fare *Truckstop Women* (1974) and *CB Hustlers* (1976), the device was used by prostitutes to solicit clients, while *White Line Fever* (1975) provided the rough blueprint for the genre of CB-films (to use the term loosely). In the film, Jan Michael Vincent takes on a corrupt trucking company, warning his adversaries of an impending attack via his CB. 1977 saw a number of CB films, including *The Great Smokey Roadblock,* starring an aging Henry Fonda, and *Breaker, Breaker,* a routine action film with a young Chuck Norris. *Smokey and the Bandit* from that same year remains the most well known of this genre. The film stars Burt Reynolds as a legendary trucker who must race a shipment of beer cross-country. CB radio is featured so prominently in the movie that some current viewers might require a jargon guide to keep up. Jonathan Demme's 1977 dramatic comedy *Citizens Band* also put this technology at the center of its overlapping subplots, though this work does not otherwise conform to the conventions of the CB films of the decade. The film was not a box-office success, though it does offers insightful observations about the impersonal, anonymous nature of communicating via CB. In the film, "Spider" monitors the local airwaves in order to aid those in need, and even takes it upon himself to silence inappropriate radio chatter.

Songwriters also jumped on the CB bandwagon, using the already-established genre of truck-driving songs as a convenient starting point. In 1976, the same year in which more than 10 million CB radios were sold, three hit country songs gave a central role to the technology: "Convoy," "Teddy Bear," and "White Knight." The first of these songs captured the public's attention at the height of the craze, reaching number one on *Billboard*'s pop and country charts, and even found success in other countries. In an attempt to duplicate the box office success of *Smokey and the Bandit,* British studio EMI translated the song to the big screen in 1978. The narratives of the song and film differ slightly, though both tell essentially the same story—a convoy of renegade truckers use CB radios in a cat-and-mouse cross-country race against the police. Kris Kristofferson stars as "Rubber Duck," the leader of the convoy, while famed western director Samuel Peckinpah attempted to inject some serious intellectual substance to this formulaic tale (though he was ultimately unsuccessful).

In contrast to the quote at the outset of this chapter, the sudden popularity of CB did warrant attention from several academics. In particular, two 1979 articles from the *Journal of Popular Culture* attached great significance to the phenomenon. In a study of CB lingo, Ramsey wrote:

> The mass proliferation of the Citizens Band radio must rank as the foremost electronic-media phenomenon of the 1970s, rivaling the impact which other technological innovations—the telephone and the television—have had on American life.[17]

On a similar note, Dannefer and Poushinksy wrote in the same journal:

> Thus the full implications of CB radio for the next few decades are not yet fully discernible. It seems clear, however, that this medium will have a major and irreversible impact on modern experience and modern social formations.[18]

While CB was unquestionably a major media phenomenon from this decade, both of these statements exaggerated the impact of the new technology.

In 1977, in order to ease the ever-worsening problem of congestion, the FCC expanded citizens band to include 17 new channels, though the decision had the unintended effect of dampening the market. Merchants slashed prices on the inventory of 23-channel radios for the 1976 holiday season, expecting consumers to snap up the new 40-channels units. The new market failed to appear, due to high prices for the new equipment.[19] In 1978, the FCC abandoned licensing requirements completely, though both of these regulatory shifts appear to have come too late. Sales of CB radios continued to drop. Exacerbating the problem was the increased level of sunspots in the later part of the decade; the resulting atmospheric conditions caused some normal CB communications to travel far beyond the normal transmission range.[20] When the FCC established the original 23-channel band in the late 1950s, radio amateurs warned of the interference triggered by the 11-year sunspot cycle. It was unfortunate timing that one of these peaks occurred during CB's popularity.[21]

The service persists, and remains in use today, though it once again is dominated by the trucking industry. The most common reason cited for this decline, beyond the simple fact that its novelty had worn off, was the sheer overcrowding of the channels.[22] With so many users talking at once, any kind sustained conversation was impossible and many casual users decided it wasn't worth all the effort.

Rule-Breaking Rebels

While the popular perception of CB radio as a trucker's tool may seem logical, given the significance of truckers to its initial popularity, most of those who flocked to the fad in subsequent years were not involved in this profession. Within the popular press, however, and the rash of films and novelty songs, CB was continually linked to truck drivers and portrayed as a means for evading law enforcement, orchestrating political protests, and for subverting social mores in general. CB radio was presented as something of a liminal space, distinct from standard society, where transgressions were acceptable.[23]

The title of a 1975 book succinctly summed up the romantic appeal of this particular profession—*Trucker: A Portrait of the Last American Cowboy*.[24] Shane Hamilton's history of the trucking industry in the 20th century argues that these drivers "reinvigorated a form of the agrarian myth."[25] The original version of this myth suggested that farmers, independently working and living off the land, were the true representatives of American concepts of liberty and democracy. With the decline of small farms in the 20th century, and the concurrent growth of urban centers, the long-haul truck driver inherited this myth, especially those independent operators who did not work for large companies. Working for a large corporation was seen as dehumanizing, while those who controlled their own destiny and owned their own trucks were the "real men" of the era. Many Southerners turned to the trucking profession as the number of small family-operated farms declined, and the trucking lifestyle was thus strongly associated with this geographic region. In 1976, for example, *Time* magazine described the nature of CB jargon "and the *Arkahoma* accent in which it is invariably delivered no matter where in the U.S."[26]

These independent truckers, who represented only a minority of drivers, were central to the volatile protests of 1973–74, and were glorified in films. *White Line Fever* opens with a faux news clip about the plight of independent truckers, before moving to a routine "little-man-versus-corrupt-company" plot. The political angle returns at the end of the film with all of the truckers in Arizona pledging to go on strike in solidarity with the hero. The later films *Steel Cowboy* (1978) and *Convoy* (1978) also made a point of emphasizing the independent nature of their protagonists. In the second of these works, the massive convoy of trucks lead by "Rubber Duck" attracts so much press attention that he is anointed the figurehead of a nascent political movement. The aims of this political movement are not clear, though presumably its emphasis was on a celebration of individuality, in the face of creeping government power that threatened to stifle American freedom.

This depiction of independent operators as the leaders of a growing anti-government movement tied directly into the depiction of CB radio as an unfiltered tool for the common man to air his opinions. "TV is, after all, a nonparticipant pastime," wrote *Time* magazine in 1976. "CB radio, by contrast, is a two-way medium that enables everyman to write his own script."[27] Many of the CB radio guides and handbooks from the decade made even more explicit statements in this vein, painting television as a corporate-controlled, one-way instrument of propaganda. One 1977 guidebook stated:

> CB is the door to a new world of communication that has no experts and no authorities, and that promises to become a more and more important part of the culture....On CB there is no Big Brother...there is only Good Buddy (so far).[28]

In contrast to the political, populist tone frequently associated with CB, it must be noted that the most successful film to exploit the phenomenon, *Smokey and the Bandit,* deviated from this message. Unlike the protagonists of the previously cited films, Burt Reynolds is not under any financial duress and engages in a wild cross-country trek simply in hopes of winning a lucrative bet. The Bandit and his partner Snowman rely heavily on their CB radios to evade police as they haul their illegal shipment. In this fictional world, the Bandit is something of a Robin Hood character, so famous and beloved that fellow truckers aid in his escapade by broadcasting the location of nearby police.

Police and Prostitution

According to a 1978 study, avoiding police was not the main use for CB, at least for those who participated in the academic study.[29] By contrast, many popular culture representations of CB focused on its ability to help drivers evade authorities. The 1973 song "Smokey, Trucks, and CB Radios" characterized the device as a way to avoid speeding tickets, among other hazards of the road. According to some background chatter in the song, presented in a static-filled voice meant to imitate a radio transmission, one trucker warns that "Smokey got him an 18-wheeler up there...that boy must not have a CB radio." As a further indication of the "evading police" theme in popular culture, one only has to look at any of the CB dictionaries of the period, or the many that have been posted online. While "Smokey" and "bear" were the most popular slang terms for police, there is in fact an extensive vocabulary to distinguish local vs. state police, police on motorcycles, those in helicopters, those with radar guns, etc.

In addition to helping drivers avoid speeding tickets (or avoiding trucking weigh stations), there was another illegal activity that was frequently linked with CB radio—prostitution. The aforementioned *Truckstop Women* and *CB Hustlers* used prostitution as central plot elements, while in *The Great Smokey Roadblock*, Henry Fonda must transport a group of "working girls" cross-country. In one of the subplots, of *Citizens Band*, a prostitute known as Hot Coffee purchases a mobile home (and CB) to update her operation. Articles in *Time* magazine and the *New York Times* also linked prostitution to CB radios; in one such article, the spread of "CB-VD" is singled out. This phrase denoted sexual diseases spread through contacts that had been initiated through CB radio, and "the carriers are almost impossible to identify because they introduce themselves with the CB handles rather than their real names."[30]

"On the Internet, Nobody Knows You're a Dog"

In 1993, a few years before most Americans obtained regular Internet access, a now famous cartoon appeared in the *New Yorker*, with the title of this section as the caption.[31] This saying has become something of a cliché over time, emphasizing the anonymous nature of online interactions in which identity is more of a performance act than stable entity. A similar phenomenon occurred with CB radio in relation to the use of self-selected handles. In 1976, a sociologist said the technology allowed the presentation of a false self "like the traveling salesman who drops into a singles bar and says he's the president of his company, a person can project on the air anything he wants to be."[32] *Time* magazine took notice of the phenomenon in 1973, and listed some of the more unusual nicknames that could be heard over the air, including Number One Nose Picker, Popper Stopper, Bootlegger, Mule Skinner, and Granny Go Go.[33]

This particular characteristic of CB communication was recognized immediately, and next to the "rebel-trucker-evading-police" motif was the most prevalent theme for popular depictions of the technology. The inability to connect the voice from a CB to a particular individual appeared in numerous films. In *Breaker, Breaker*, for example, corrupt police broadcast false messages to unsuspecting truckers, diverting them to police traps. In *Smokey and the Bandit*, the villainous sheriff (played by Jackie Gleason) comes face to face with the Bandit in a diner, though is unaware of his true identity. The Bandit, however, is fully aware of who he is talking to, using the brief moment to mock his adversary. A reporter attempting to uncover the illegal prostitution ring in *CB Hustlers* also struggles to match CB voices to specific individuals, though *Citizens Band* was the work to most fully explore the phenomenon of anonymous or outright deceptive communication. In the film, a number of individuals deliberately hide their identities with handles, including the main character's girlfriend, who calls herself "Electra" while engaging in provocative "radio sex" with unknown men. Upon discovering the truth, the main character states quite directly: "Everybody in this town is somebody they're not supposed to be."

Among the novelty songs of the decade, "The White Knight" also revolves around an on-air deception. A trucker, known only by his handle, speaking in a distinctly Southern accent, instructs the narrator to put "the pedal to the metal and let it roar." The unsuspecting narrator believes the White Knight is driving "front door," or leading the convoy of trucks, and therefore can see what is coming up before the others. At the end of the song, it is revealed that a local policeman has set a clever trap and apprehends the hapless narrator for speeding. A lesser-known, though even more bizarre, song also released in 1976 presents a similar scenario. "CB Savage" tells the story of a policeman who adopts the kind of high-pitched, exaggerated lisping voice used to mock homosexuals. He broadcasts sexual propositions to nearby truck drivers, who are so distracted that they fail to notice a speed-trap. In this song, the common CB phrase "come on," which is a standard request for a response, acquires a more provocative meaning. More so than other artifacts from the era, the song exemplifies the heterosexual masculinity associated with the use of CB, with homosexuality and law enforcement conflated as that which must be avoided.

Chaos and Congestion

The growing popularity of the CB craze, which caught the FCC off guard, led to a severe overcrowding of the channels. This situation was ultimately blamed for the rapid decline of the fad, though even before the end of the decade, the congested airwaves and CB's tendency to bleed into other portions of the spectrum were identified as problems. A 1974 newspaper article on the situation described CB as a "jumbled morass of chitchat [that] is virtually impossible to police."[34] Two years later, the FCC was flooded with "CB horror stories," including "reports of automatic garage doors responding to the spill-over energy from citizens band transmissions, phonographs and public-address systems picking up from the CB sets, and television pictures suffering severe disturbances."[35] The problem was said to be particularly bad in the Los Angeles area, where the volume of users clamoring for airspace and a handful of malicious individuals threatened to drown out any intelligible conversation. Such complaints about interference recall the criticism of radio amateurs during the original era of wireless telegraphy, when mischievous operators would broadcast false distress signals or otherwise disrupt normal operations. In order to maintain a sense of order on the air, the FCC shunted amateurs to the low-end of the spectrum, where they wouldn't interfere with government or commercial wireless stations. In the 1970s, by contrast, the FCC sought a different solution to the problem of airwave congestion and instead opened up new channels for CB users.

CB transmitters were ostensibly designed only for short-range communication, though a variety of devices allowed unscrupulous users to greatly boost their power. If you were "running power," according to the jargon, you could "walk on" other users and dominate a single channel in a vast geographic area. Malicious users who did not respect the etiquette of the airwaves were dubbed "rachet jaws" or "motor mouths," and attracted much wrath from other CB operators. In October 1976, the *New York Times* reported one such instance that had deadly consequences.[36] According to this report, "Blue Goose" of Fort Worth, Texas was dominating the local channel six, an act that so angered "Dirty Bird" that a gunfight ensued.

The plot of *Citizens Band* deals with this exact problem of airwave congestion. The character "Spider" devotes his time to monitoring the CB radio, listening for anyone in need of assistance, and is frustrated by the inane chatter, including racist rants and bombastic evangelizing, that is blocking out certain frequencies. In response, he and a friend take it upon themselves to become "airwave vigilantes," dismantling antennas and smashing the CB radios of those guilty parties.

Conclusion

While CB radio differs in a significant way from current communication technologies (specifically the crowding of all users into such a limited portion of the airwave spectrum), the ways in which it was conceptualized do share much in common with present phenomena. The mobile social network of CB, for example, was depicted as something unruly and not subject to control, with anonymous users able to evade law enforcement, and this anti-authoritarian aura has been transferred to newer forms of communication. Modern social networks are the platform for not only the most mundane of conversations, but also for salacious or provocative activities that circumvent regulations established in the past. And, it should also be noted that prostitutes have again exploited a new form of electronic communication as part of plying their trade.

The link between CB and protests must also be acknowledged, especially since at the time of this writing, news headlines continue to trumpet the role of social networks in organizing the volatile protests in several Middle Eastern countries. Setting aside the question as to whether or not such snap judgments are accurate (or simply manifestations of a determinist philosophy), CB established a precedent for glorifying social networks as a "method of protest." CB first came to the attention of many individuals through its use in several highly publicized protests orchestrated by truck drivers, and in the public consciousness, this link between rebellion and decentralized communication persisted.

Another parallel between the CB chatter of the past and online communication of today is the performative nature of such communication. With CB, users chose their own handles to distinguish themselves and engaged in a linguistic wordplay; the specific words used to convey a message were just as important as the content of the message itself. In today's environment, members of online social networks can choose whatever name or image they desire, with some individuals displaying great creativity in such matters. The online world is a stage on which our identities are continually performed. A parallel between the colorful CB jargon and text messaging and Twitter can also be observed. Certain abbreviations are used, or specific phrases, even when not mandated by character limits but rather in a playful nature reminiscent of CB jargon.

While neither the extreme spectrum scarcity of CB or the atmospheric interference issue plague modern social networks, the continued laments over "the morass of chitchat" foreshadowed contemporary critiques. Complaints over mobile phone usage in public locations, for example, have become a staple of modern life with certain locations, including theatres and public transportation, perceived as particularly inappropriate. Criticisms of the flood of useless conversation are even greater for social networks. A study of Twitter, for example, identified "information overload" as one of the recurring themes within mainstream press coverage, with tweets about food singled out as particularly useless.[37] While proponents celebrate the potential benefits of mobile phones and social networks, these recurring critiques indicate that some individuals also lament the lack of gate-keepers who might ensure that only "valid" messages are communicated.

But perhaps the greatest similarity between CB radios and online social networks is the hyperbolic praise that each has generated from academics and journalists. The previously cited statements regarding CB radio, which was said to have a social transformative power equaling the telephone and television, cannot stand the test of time. The critique that American media industries exert an undue amount of influence upon society is an old saw; the monopolistic power of Western Union over the telegraph, for example, was targeted well over one hundred years ago. Given the longevity and persistence of this particular critique within American society, it is not surprising that each new innovation in electronic communications, from the telephone to radio to television to cable television to the Internet and now to social networks, has been trumpeted as "giving voice to the common man." And, while the problem of interference that ultimately slowed the growth of CB radio is not likely to reappear with online social networks, this retrospection into the fad should give one pause regarding current utopian hopes for electronic media. The promised media revolutions of the past failed to trigger as much social change as their proponents hoped, and the social networks of today may very well become mundane parts of the media landscape of the future rather than the revolutionary tools that the popular media often frames them to be.

Notes

1. Jeremy Packer, "Mobile Communications and Governing the Mobile: CBs and Truckers," *The Communication Review* 5, no. 1 (2002): 41.
2. Though CB radio has not been well studied in the past two decades, the technology did inspire several academic studies in the late 1970s and early 80s. A 1977 (no. 3) issue of the *Journal of Communication* featured three articles on the topic: Carolyn Marvin and Quentin J. Schulze, "The First Thirty Years"; Jon T. Powell and Donald Ary, "Communication without Commitment"; W. Dale Dannefer and Nicholar Poushinsky, "Language and Community." Other examples of previous academic studies include Richard David Ramsey, "The People Versus Smokey Bear: Metaphor, Argot, and CB Radio," *Journal of Popular Culture* 13 (1979); W. Dale Dannefer and Nicholas Poushinsky, "The C.B. Phenomenon, A Sociological Appraisal," *Journal of Popular Culture* 12, no. 4 (Spring 1979). Two recent studies of the phenomenon include Angela M. Blake, "An Audible Sense of Order: Race, Fear, and CB Radio on Los Angeles Freeways in the 1970s," in *Sound in the Age of Mechanical Reproduction*, ed. David Suisman and Susan Strasser (Philadelphia: University of Pennsylvania Press, 2010): 159–178; also Tyler Watts and Jared Barton, "I Can't Drive 55: The Economics of the CB Radio Phenomenon," *The Independent Review* 15, no. 3 (2011).
3. Jennifer S. Light, "Facsimile: A Forgotten 'New Medium' from the 20th Century," *New Media & Society* 8, no. 3 (2006): 372.
4. Gerard Goggin, *Global Mobile Media* (New York: Routledge, 2011), 117.
5. Noah Arceneaux, "The World Is a Phone Booth: The American Response to Mobile Phones, 1981–2000," *Convergence* 11, no. 2 (Summer 2005); see also Sergio Rizzo, "The Promise of Cell Phones: From People Power to Technological Nanny," *Convergence* 14, no. 2 (2008).
6. Packer, "Mobile Communications and Governing the Mobile: CBs and Truckers."
7. Harold R. Kerbo, Karrie Marshall, and Philip Holley, "Reestablishing 'Gemeinschaft'?: An Examination of the CB Radio Fad," *Urban Life* 7, no. 3 (October 1978); see also Powell and Ary, "Communication without Commitment."
8. Carolyn Marvin and Quentin J. Schulze, "The First Thirty Years." Much of the information in this section, unless otherwise cited, derives from this source.
9. Michel de Certeau, *The Practices of Everyday Life* (Berkeley: University of California Press, 1984).

10. Jane Stern, *Trucker: A Portrait of the Last American Cowboy* (New York: McGraw-Hill, 1975), 156.
11. "The Truckers Hooked Us on CB" (quoting the trucker "Ribbon Wrangler"), *Country Music*, December 1976, 50. This article was part of a "Special CB Section."
12. Blake, "An Audible Sense of Order," 162.
13. W. Dale Dannefer and Nicholas Poushinsky, "The C.B. Phenomenon, A Sociological Appraisal," *Journal of Popular Culture* 12, no. 4 (Spring 1979): 611.
14. These numbers based on a chart of CB imports in Watts and Barton, "I Can't Drive 55": 389.
15. Watts and Barton, "I Can't Drive 55": 389.
16. Beth Ann Krier, "The Radio That Ate Los Angeles," *Los Angeles Times*, May 2, 1976, H1.
17. Richard David Ramsey, "The People Versus Smokey Bear: Metaphor, Argot, and CB Radio," *Journal of Popular Culture* 13 (1979): 340.
18. W. Dale Dannefer and Nicholas Poushinsky, "The C.B. Phenomenon, a Sociological Appraisal," *Journal of Popular Culture* 12, no. 4 (Spring 1979): 619.
19. Lynn Langway, "10–4 for CB," *Newsweek*, January 24, 1977, 68.
20. "Scientist Says CB Radios Face Sunspot Interference," *New York Times*, July 11, 1976, 28.
21. Marvin and Schultze, "The First Thirty Years": 100.
22. For one such explanation, see Jesse Walker, *Rebels on the Air: An Alternative History of Radio in America* (New York: New York University Press, 2001), 274.
23. Victor Turner, *Drama, Fields and Metaphors: Symbolic Action in Human Societies* (Ithaca, NY: Cornell University Press, 1974).
24. Stern, *Trucker: A Portrait of the Last American Cowboy*.
25. Shane Hamilton, *Trucking Country: The Road to America's Wal-Mart Economy* (Princeton, NJ: Princeton University Press, 2008), 189.
26. "Modern Living: The Bodacious New World of C.B.," *Time*, May 10, 1976, 78–79.
27. Ibid.
28. Original statement from *The All New Fact-Packed 1977/78 CB Guide* by Schlossberg and Brockman, quoted in Packer, "Mobile Communications and Governing the Mobile": 43.
29. Kerbo, Marshall, and Holley, "Reestablishing 'Gemeinschaft'?": 343–344.
30. "Pranks, Theft, Disease and Other Fallout from the CB Boom," *New York Times*, May 22, 1978, B1.
31. Original cartoon from *The New Yorker*, July 5, 1993, 61.
32. "Modern Living: The Bodacious New World of C.B.," *Time*.
33. "Modern Living: Voices on the Road," *Time*, December 3, 1973, 80.
34. "Usefulness of Citizens' Band Radio Periled," *Los Angeles Times*, October 18, 1974, B8.
35. "CB Interference Brings FCC Horror Stories," *New York Times*, July 10, 1976, 49.
36. "Blue Goose Is Dead and Dirty Bird Is Hurt in CB Radio Dispute," *New York Times*, October 16, 1976, 34.
37. Noah Arceneaux and Amy Schmitz Weiss, "Seems Stupid Until You Try It: Press Coverage of Twitter, 2006–09," *New Media & Society* 12, no. 8 (December 2010).

5

A Brief History of U.S. Mobile Spectrum

Thomas W. Hazlett

When cellular telephone technology was first crafted, at AT&T's Bell Labs in 1947,[1] it seemed like a very good idea to the scientists. Mobile telephone phone service (MTS), with huge transceivers housed in automobiles, had been launched in 1946. Not only was the equipment unwieldy and expensive, networks had severe capacity constraints, featuring—for example—only six channels (hosting just six conversations) throughout all of St. Louis. Even at sky-high prices, demand well exceeded supply. Long waiting lists developed to obtain a subscription, which was itself only a hunting license, as users were continually queued up, waiting for an open channel. Cellular engineering—allowing spectrum to be re-used from "cell" to "cell," with low-powered radios (handsets) linking to each cell's local "base station"— was an ingenious way to add channels and greatly expand service.[2]

The Federal Communications Commission was not so impressed. While the first application to the FCC for cellphone service was in 1958, progress was slow. The request targeted the UHF band, which had been designated for over-the-air television service. Broadcast TV stations saw a reallocation of "their" airwaves for the use of other industries as an appropriation. Perhaps AT&T, with a monopoly position in (most) local and (all) long distance phone services in the U.S., did not sufficiently press the issue. Or, as some have argued, "the agency tasked with controlling [spectrum] use—the FCC—is historically one of the more dysfunctional bureaucracies in Washington."[3] Until 1968, the Commission refused to act.

The Long Wait for Cellular

While the Commission had officially opened a proceeding to consider the application to authorize cellular telephony, this was only the beginning of a formal rule-making which would consume nearly two full decades. During those years the FCC would re-allocate UHF spectrum from television (specifically, channels 70–83) to cellular, craft regulations governing the new voice services, and then assign cellular licenses to phone companies (1983–89). Counting this as merely a 10-year regulatory lag cost the U.S. economy an estimated $86 billion.[4] What took so long?

The FCC has been called an "attractive nuisance" due to its structure and the legal and political realities surrounding it. The agency is bound by administrative law to make deliberate choices according to "public interest, convenience or necessity"—a purposely vague standard. The agency has lots of wiggle room to shape its decisions according to the arguments that competing interest groups offer. These arguments do not have to make a lot of sense, but simply have some "rational connection" to the "public interest." For instance, the FCC adopted a plan for broadcast television (in 1952) that had the predictable effect of restricting the vast majority of American viewers to no more than 3 program choices—those offered by ABC, CBS, or NBC. This, despite setting aside the bandwidth for 82 TV channels nationwide. Competition was simply not considered an important goal by the Commission, a viewpoint that the three incumbent TV networks agreed with heartily. As a result, the vast majority of TV channels have been "white spaces," vacant spectrum producing no value for society. This is true even now, generations of technology—and one analog-to-digital TV transition—later.

Authorized by Congress in the 1934 Communications Act, the FCC is a five member body. Commissioners are nominated by the President and confirmed by the U.S. Senate. No more than three members can be from the President's own political party. The agency is charged with regulating the use of radio waves in the United States. (Government spectrum use, however, is coordinated by the National Telecommunications & Information Administration at the U.S. Department of Commerce.)

In licensing airwave use,[5] the Commission operates passively. That is, it typically waits for parties to file petitions or applications, rather than organizing wireless markets based on its own plans. There is a very good reason for this: almost all the relevant information about consumer demand, wireless technology, investment opportunities and business strategies is dispersed throughout the economy. The government regulator must discover what possibilities exist to create valuable wireless services through a revelation process. That occurs when firms develop technologies and then go to the FCC for permission to deploy them.

While this brings the regulator potentially important data, it also gives many others the same public information. Competing interests will make similar requests, or file petitions to oppose any concessions to their rival. Incumbent licensees have sharp interests in fending off entry both to deter competition and to mitigate potential costs (as with radio interference spillovers inflicted on their customers). TV broadcasters long opposed the use of UHF radio waves for new cellular services, despite the fact that it would have imperceptible impacts on broadcast reception.

While established interests had the ability to delay a cellular allocation, it was difficult for cellular's backers to amass much of a counter-attack. Uncertainty played a role. Many experts insisted that cellular telephony would not amount to very much. McKenzie & Co. had issued a projection for AT&T in 1980: by the year 2000, there would be only 900,000 cellular phone subscribers. The forecast was off by over two orders of magnitude; in 2000, subscribership exceeded 109 million.[6] The conventional wisdom led FCC regulators not to take the emerging technology seriously. For instance, "Commissioner Robert E. Lee had opposed a large allocation of radio spectrum, calling it a 'frivolous use' of the public airwaves to provide 'each automobile owner another status symbol—a telephone for the family car.'"[7]

Throughout the 1970s and early 1980s the FCC argued the matter. By 1974, regulators had tentatively decided to issue one cellular license in each market, and to issue it to an AT&T-affiliated phone carrier. The presumption was that telecommunications was monopolistic, and that—given economies of scale—market competition would be untenable. Indeed, an independent monopoly would be financially inviable; it would need the support of the local fixed-line phone carrier to survive. The initial choice was reconsidered, and a then-radical scheme was adopted: the FCC would authorize two cellular networks per market. One would be operated by the local fixed-line phone company, the other by an independent wireless phone carrier. It was a watershed moment in telecommunications policy, and some suspected that the experiment in head-to-head network rivalry would prove unworkable.

Finally, the FCC charged forward, issuing two licenses in each of 734 local markets. This was an extraordinarily 'atomistic' franchising scheme. When other countries began allocating wireless licenses, the standard was to issue nationwide authorizations. To award some 1,468 permits by the old method of "beauty contests," where rival groups of lawyers concocted "public interests" arguments about why their clients should receive—at zero price, licenses worth millions—was administratively untenable. While FCC and Executive Branch officials had, since the Carter Administration, requested that Congress allow the agency to auction licenses, such reforms were bitterly resisted by key committee chairs in Congress. (These policy makers enjoyed the power gained by being in the middle, as legislative oversight of the FCC, of brutal political fights for lucrative licenses.)

But in 1981 Congress did authorize the use of lotteries, a curiously perverse assignment tool which dispenses with the fig leaf of "public interest" awards but refuses to let the U.S. Treasury collect the bids. The licenses are distributed at random, and bids are then traded in secondary markets by lucky winners and their lawyers. Lotteries were used in two big assignment rounds. Major market licenses (305 markets) were issued 1984–86; small market licenses (428 markets) were assigned 1987–88. Hundreds of transactions in secondary markets then ensued, networks were stitched together, and a new industry was launched.

The Early Cellular Market

In the mid-1980s, cellular services finally hit the market. Phones were expensive—about $2,000—and monthly service, too—about $100 or more.[8] Networks were crude; with 734 market areas licensed, services were highly fragmented. Roaming was difficult and costly. And "AMPS" analog technology, mandated for use by the FCC, was already becoming obsolete.[9]

But cellular was nonetheless proving popular. Investors noticed that the pessimistic forecasts were being blown away. Cellular licenses became hot properties. Prices leapt from about $20 "per pop" (per person in the licensed area) in prime urban/suburban markets in 1985, to over $200 in 1989.[10] As of 1990, the U.S. Department of Commerce estimated that the value of all cellular licenses totaled about $100 billion.[11]

This marketplace enthusiasm signaled that the initial spectrum allocation had been exceedingly parsimonious. Consumers were embracing the new services, and additional bandwidth would make it possible to deliver them more efficiently. Economists would point out that this did not imply a "spectrum shortage." Prices were capable of limiting network access and mitigating congestion. The inefficiency was that those prices were far higher than necessary, much above what they would have obtained had more spectrum—or better technologies, beyond AMPS—been available for deployment. Resources that were as plentiful as land in Kansas were being treated as though they were prime acreage in New York City. That formed a regulatory bottleneck.

The history of mobile services in the U.S. reflects a distinct pattern. Long delays block economic development; when the bottleneck loosens, permitting dollops of new bandwidth to go to entrepreneurs, frenetic economic activity ensues—new services arise, prices fall, and mobile use skyrockets.

There are two sets of lags: one associated with spectrum allocations and license assignments, the other with actual network deployments. The first set is uneconomic. There is no social gain derived from delaying private investors from utilizing bandwidth to provide services that they believe customers would like to pay for. The second set of lags is, conversely, explained by actual cost constraints. It does not make sense to build a house—or a mobile network—in a day. Indeed, it is often sensible to wait a bit to build (or upgrade) a network, as technologies develop to improve performance.

The key difference between the situations is that policy makers who delay valuable services pay no price for the imposition they cause. The situation is precisely reversed for private firms. Just as the owner of an oil well attempts to pump at a pace that maximizes her wealth, the licensee delays spectrum deployment only when the gains from pumping a bit slower are anticipated to offset the costs.[12]

A Sneak Attack on the Cellular Duopoly

During the initial regulatory lags in cellular, the lack of licensed spectrum preempted the development of any commercial operations until 1983. Then, with the assignment of two cellular licenses, each allocated 25 MHz, in markets nationwide, the service came to life. Networks were built and customer demand was surprisingly robust. Wireless carriers, who had initially planned to make money by selling expensive handsets, giving away cheap telephone access,[13] adjusted their business models to focus on selling service contracts.

Morgan O'Brien, a former FCC lawyer who had dealt with the regulation of taxi-dispatch licenses, saw that "specialized mobile radio" (SMR) permits authorizing mobile dispatch services were allotted UHF spectrum adjacent to cellular's allocation. He realized that they could easily provide far more valuable services *if* the licensees were permitted to compete for cellular customers. It was a big "if." Under the terms of the SMR license, firms were not allowed to directly compete with mobile carriers. But that might change were the FCC to see the "public interest" as thereby served.

O'Brien raised some capital and quietly began buying SMR license rights in secondary markets. He then prepared an application for a regulatory waiver, permitting his company—Fleet Call—to digitize its dispatch service, and use the additional capacity thereby generated to provide additional services, including mobile calls to other fixed and mobile phones. Obtaining the FCC waiver was bitterly opposed by incumbent cellular operators, O'Brien became the exception proving the rule. With extraordinary inside knowledge of Commission procedures, and a fair amount of entreprenuerial pluck, Fleet Call was permitted to enter the cellular business. The firm, which later became Nextel,[14] provided new rivalry, a new wireless technology (Motorola's i-DEN), and a new service (push-to-talk, a peer-to-peer intercom service for fleet customers). Three national carriers were now in operation, one a total surprise to the regulators. When the Nextel network was acquired by Sprint in 2005, it served 15 million subscribers and sold for $35 billion.[15]

PCS Spectrum to the Rescue

Prices were still high, however; over 50 cents per minute of use through 1995. Network access was consequently truncated. Lower prices would trigger the consumption of far greater quantities—and the trigger for lower prices was more spectrum. In 1994, the Federal Communications Commission began to assign PCS ("personal communications services") authorizations by competitive bidding. The task took more than a decade, with the last auctions completed in 2005.[16] The first two, PCS-A and PCS-B, were assigned expeditiously, however. Each was allocated 30 MHz; when they were distributed to carriers in 1995, it doubled the mobile bandwidth available.

The new licenses allowed operators to choose the digital standard of their choice. This freedom had been extended to cellular licensees in 1988, but the liberalization had come after the big-city systems, licensed by 1986, had been constructed with the AMPS mandate. The new PCS licensees forced the issue. Operators quickly selected standards, choosing among GSM (widely used in Europe), TDMA, and CDMA (a data-friendly technology developed by San Diego-based Qualcomm), and built modern digital systems.

The infusion of bandwidth made a huge impact. Faced with new rivals using advanced technologies and multi-feature handsets, incumbents found themselves facing dangerous rivals. They began the expensive task of refitting analog networks for digital technologies, overlaying new networks on existing infrastructure. Customers were given new digital handsets and carefully migrated to emerging digital networks, sharing licensed bandwidth with the carrier's analog subscribers. Retail prices began crashing.

It is interesting to note that the PCS allocation saw regulators place little emphasis on the importance of new competition. Rather, they considered the allocation to be a way to accommodate a new technology. The buzz was that PCS would allow customers to travel with a portable device, plugging it in at home, then using it in the car, and then the office. One telephone number—which people would keep from birth to death—would now be able to reach them no matter where they were located. Users would be continuously connected and always in touch. This was the essence of a speech given hundreds of times by FCC officials in the 1990–94 period.[17]

It need not be argued that the vision was faulty. Today's business card typically carries multiple phone numbers (not to mention email addresses, URLs, and Twitter I.D.s), and people are harder to reach on the telephone than ever before. But PCS was an enormous success simply in infusing the market with additional bandwidth. This allowed new forms of competitive enterprise to emerge.

The key disruption came courtesy of a marketing innovation by AT&T Wireless. In May 1998 it introduced Digital One Rate. This bundle featured a bucket of monthly minutes lacking restrictions as to time of day or (within the U.S.) geography. Coast-to-coast calls cost what phoning across the street did. And now those calls were effectively "free" up to the size of the bucket. A dagger was aimed at the heart of long-distance carriers and rival mobile carriers. Customers flocked to the bundle, plunging the dagger.[19] Within months all of AT&T Wireless' major competitors had adopted similar pricing plans of their own. Prices, in average revenue per minute of use (MOU),[20] went into free fall. By 2000, they were at 20¢; by 2003, 10¢; by 2007, 5¢. See Figure 1.

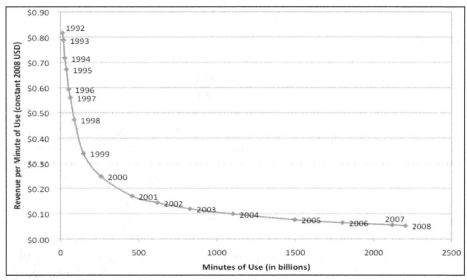

Fig. 1. Annual Mean Prices and Mobile Voice Minutes of Use (2008 Dollars)[18]

A Decade Long Spectrum Drought

But while the build-out of 2G (PCS) networks led to an explosion in voice usage, the next generation of technology, already visible, was being held up by regulators. Specifically, 3G mobile systems offered broadband data connections. While European Union countries, including England, Germany, Switzerland, Italy, France, Austria, Belgium, Greece, and Denmark, issued licenses for 3G networks in 2000–2001, the U.S. did not. To its credit, the U.S. maintained a liberal policy with respect to existing licenses: operators were free to upgrade, if they could do so in the frequency spaces allotted with the (1G, 2G) licenses already issued. But bandwidth was needlessly constrained.

From 1994, when the 120 MHz PCS allocation was made by the FCC, until 2006, when the FCC auctioned licenses allocated 90 MHz of Advanced Wireless Services (AWS) licenses, the Commission failed to make more radio spectrum available to the mobile market. This was due in large measure to the problems the FCC encountered in instituting license auctions, authorized by Congress in 1993. While competitive bidding was—and is—a superior way to assign rights, as compared to beauty contests or lotteries, egregious errors were made in the PCS C and F block auctions. The Commission elected to help small businesses and rural telephone carriers bid against larger firms by extending such "designated entities" (DEs) special terms. These included bidding credits, where a dollar bid was recorded in the auction as $1.40, and long-term low-interest loans. If a DE won a license with a $100 million bid, it paid only interest for four years, and then paid off the loan in six years, paying only the U.S. Treasury Bond interest rate—a subsidy of at least ten percent per annum.

This proved a disaster. Firms organized specifically for the purpose of entering the FCC auctions as DEs—without assets, as required by the FCC's rules, and therefore having no collateral—bid aggressively. Prices in the A and B block auction, which ended in March 1995 and which excluded DEs, averaged about 49¢ per MHz per person ("pop," referring to the population of the licensed area). In the C block auction, where DEs participated, winning bids average about *150% more* in May 1996. See Table 1.

But the games were just beginning. Having overpaid, many of the DE winners elected to declare bankruptcy, refused to give their licenses back to the FCC, and instead asking bankrupcy courts to reduce their liabilities to the government. As bold a strategy as this sounds, the major DEs—particularly the two largest, General Wireless, Inc. and NextWave, companies that registered winning bids totaling about $1 billion and $4 billion, respectively—did this, and *won*. Most damaging was not the fact that the FCC ended up receiving less than the $10.1 billion bid for C block licenses in 1996, but that the assignment of radio spectrum was held up for years. It was not until a third and final re-auction of C block licenses in 2005 that the 1994 PCS spectrum allocation was at last implemented.

During the intervening decade policy makers were distracted by the PCS C (and F) block DE license legal skirmishing, and were confused by the conflict between auction revenues and economic efficiency. It is well-known that limiting competition between service providers after the auction will tend to maximize winning bids in the auction. A legal monopoly for mobile services would invite the highest price of all. But such an approach is penny-wise, pound foolish, an extremely expensive way to raise government revenues. To gain government funds, it imposes monopoly distortions, sacrificing the very considerable economic and social benefits of robust mobile markets.[21] The simple, reliable policy rule is that license sales are a side show; the overwhelming benefits are delivered by getting abundant bandwidth quickly into its most productive employment. This, however, escaped U.S. policy makers who—even as European countries moved major new spectrum resources into the market via 3G licenses assigned in 2000 and 2001—actively sought to delay more auctions on the grounds that bids would be low.

Table 1. Major FCC Auctions for Licenses Providing Mobile Services

Auction Name	End Date	Total Rev.	Total MHz	Avg. $/MHz/pop
Broadband PCS A & B	03/13/95	$7.7 bil.	60	0.49
Broadband PCS C	05/6/96	$10.1 bil.	30	1.26
Broadband PCS D, E, F	01/14/97	$2.5 bil.	30	0.31
Broadband PCS (re-auction)	02/15/05	$2.0 bil.	n.a.	1.05
Advanced Wireless Services	09/18/06	$13.7 bil.	90	0.51
700 MHz Band	03/18/08	$19.0 bil.	52	1.20

Source: Federal Communications Commission website; author's calculations.

Searching for Spectrum by Merger and Acquisition

Lacking new allocations, alternative measures were pursued to move 3G forward in the U.S. The first was a series of secondary market transactions, and the long-awaited re-auction of PCS C bloc licenses. These deals gave Verizon Wireless, in particular, the spectrum inputs to upgrade its CDMA network for high-speed data services, deploying the 1X EV-DO standard. By 2003, it began building out a national high-speed wireless data network.[22]

Other players were forced to try different strategies—a merger wave ensued. By acquiring another cellular network, a carrier could effectively harvest additional spectrum.[23] In 2004, Cingular, a joint venture of SBC and BellSouth, purchased AT&T Wireless for $41 billion.[24] Almost immediately, the newly merged firm (which later chose to call itself AT&T) announced that it would begin upgrading its wireless network to supply 3G services.[25] The correlation between the two events was causative, according to Cingular.[26] In the following year, the third and fifth largest national networks merged, Sprint acquiring Nextel for $35 billion. The new Sprint-Nextel quickly built out its CDMA 3G network, offering service beginning in late 2005.[27] Not only did the combination help its 3G rollout,[28] but set the stage for the 4G network now in operation via partnership of Sprint and Clearwire.

The fourth largest U.S. network, T-Mobile, found no such acquisitions in the market, and was forced to delay its 3G upgrade plans.[29] This was deleterious to the operator, which had been an early partner of Research in Motion, featuring its BlackBerry smartphone and helping it develop almost a cult-following among "road warriors" in the business community. Faced with extremely limited bandwidth, the carrier was forced to wait for regulators to make new licenses available to mobile operators. When it finally did, in the Advanced Wireless Services (AWS) auction held in August-September 2006, T-Mobile bid aggressively. Its purchase of licenses allocated about 27 MHz nationwide cost it $4.2 billion, the most bid by any participant.[30] It quickly moved to play catch-up with its rivals, announcing that it would invest an additional $2.6 billion in network upgrades to enable nationwide 3G coverage.[31] It began its rollout of services in New York City in March 2008, utilizing the AWS spectrum.[32] Other smaller carriers like Leap[33] and MetroPCS[34] also benefited from the release of additional bandwidth, and built-out advanced networks with the new spectrum.

With these twists and turns on the spectrum front, advanced wireless Internet access service began to take off. With the very first customers showing up in 2004 and 2005, momentum began to build. By year-end 2010, some 64 million subscribers received service in the U.S. for high-speed mobile data services, delivered to handsets or modems ("dongles") plugged into notebooks, tablets, e-readers or netbooks. As defined by the FCC, this is a residential service that delivers data downloads of at least 200 kbps. See Figure 2.

The historic pattern in U.S. spectrum allocation is distinct. Long regulatory delays are imposed. This freezes markets. When bandwidth is finally allocated to licenses and these licenses assigned to operators, a flurry of investment activity takes place. Vast infrastructures—mobile communications networks—are created, expanded and upgraded to complement (and so increase) the value of raw airwaves. Carriers then compete to supply services and applications that draw customers to their networks. Prices fall and usage skyrockets. Innovators build new devices, content and applications to create further value in the platforms available.

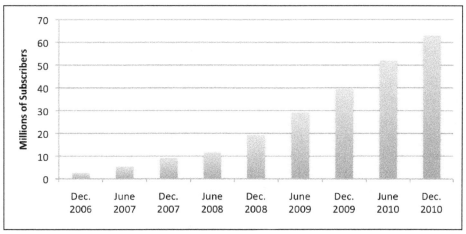

Fig. 2. U.S. Mobile High-Speed Data Subscriptions (000s)[35]

These tasks are far too complex and far-flung to be efficiently crafted by one firm. Specialization offers many benefits. Hence, mobile operators focus on the creation of networks, the aggregation of spectrum inputs, and the creation, maintenance, and management of infrastructure. Technology inputs are acquired from independent suppliers such as Nortel, Motorola, Lucent, Ericsson, or Qualcomm. Handsets are purchased from vendors such as Samsung, LG, Nokia, Kyocera, RIM, or Apple. Plug-ins and software applications are provided by a wide range of firms; they are fundamentally coordinated under basic technologies chosen by the carrier, optimized according to rules and marketing strategies put in place by non-carrier contributors such as Apple, Google (with its Android mobile operating system), RIM or Nokia. The complexity of this mix of substitutes and complements, and its spontaneously evolving nature, has given rise to the term, "mobile ecosystem."

The oxygen of this ecosystem is radio spectrum. It is the essential life-giving ingredient for mobile networks. Without bandwidth available to the market, in liberal increments and with broad rights to use frequencies efficiently, wireless traffic will not flow. Release the constraints and entrepreneurs will develop new technologies, build out infrastructure, and create the platforms on which valuable new services will travel.

This pattern is playing out again. In auctions held by the FCC in March 2008, 700 MHz bandwidth was reallocated from TV broadcasting to mobile communications. (This reallocation took, in fact, some 22 years, counting just the official FCC record.)[36] The licenses, which were primarily assigned to Verizon and AT&T, the top two mobile carriers (in subscriber size) and the top bidders in the auction, infused the market with about 70 MHz of additional bandwidth. This spectrum bump has driven the major carriers to develop and deploy next generation, 4G, LTE—"long term evolution"—wireless network upgrades. 4G launches, bringing faster speeds (up to 12 mbps downloads) and better service to millions of customers began in 2010.[37]

Smaller firms can play this game; indeed, the availability of spectrum resources is a primary enabler of competitive entry. Metro PCS is the fifth largest U.S. mobile operator, serving about 8 million customers. The firm employs a low-cost business model to supply sharply discounted packages to garner market share. For instance, its "all you can eat" voice, text, and data plans start at $40 per month, much below rival plans that (with faster networks) can cost $120. Metro PCS is using the bandwidth acquired in the 2006 AWS auction to roll-out 4G services, effectively leapfrogging 3G technology.[38] The innovation is driving new equipment into the market, as AWS frequencies (in the 1.7 GHz and 2.1 GHz bands) have not previously been incorporated in LTE designs.

Would even more bandwidth be better, yielding superior services and faster data delivery? Not much of a question, actually:

> MetroPCS warned in its SEC filing that its forthcoming LTE network might not wow users with lightning-fast downloads. "In some cases, because of the limited amount of spectrum available to us in certain metropolitan areas, we will be required to deploy LTE on 1.4 or 3 MHz channels," the carrier said. ....In comparison, Verizon Wireless plans to devote up to 10 MHz to LTE. Because MetroPCS is squeezing LTE into such a narrow spectrum channel, the carrier likely won't be able to provide data speeds beyond what are available through today's 3G networks.[39]

Put simply, more spectrum is preferred to less, all else equal. The limiting principle is supplied by opportunity cost. Where the extra spectrum being made available to the mobile market has alternative uses that yield greater value, then the "reallocation" has gone too far. The over-riding reality in spectrum allocation, however, is that there exists vast under-utilized frequency space. Given current political process constraints, we are unlikely to generally arrive at the balancing margin any time soon.[40]

The appropriate public policy approach to "reallocation of radio spectrum" is not to mandate that frequency resources be transferred from one employment (say, TV broadcasting) to another (like mobile communications). That is the administrative *diktat* approach that delivered our current problems.[41] Rather, it is to define wireless license rights liberally, permitting market competition to allocate resources. Rival mobile operators should be able to bid for spectrum resources, shifting bandwidth from one deployment to another. This process rewards firms for discovering and delivering efficient solutions for serving customer demand. When regulatory bottlenecks are removed and spectrum flows to market, prices plummet even while networks grow in quality, scope, and sophistication. The process is in full view, 1G through 4G, and continues apace today—as seen in the forming "mobile data tsunami."

The Smartphone Revolution and the Internet of Things

Since the introduction of the iconic Apple iPhone in June 2007, mobile markets have been transformed yet again. While Research in Motion's BlackBerry had years earlier offered advanced phones that accessed email and the Internet, the consumer reaction to the iPhone spurred intense competitive interest in developing such phones for mass market adoption. Apple's App Store brought hundreds of thousands of applications directly to the handset, bypassing carriers, and instigating entirely new platforms for mobile commerce and social networking. Countered by the entry of Google's Android platform, and confronted by renewed efforts from incumbent device makers RIM and Nokia, smartphones have rapidly gone from from a business tool to a popular consumer electronics gadget trending towards dominance among mobile subscribers.

Perhaps of even greater long-run significance than the smartphone revolution, however, is the emerging Internet of Things. Devices that communicate without human supervision, call them *Im*personal Communications Services, are flourishing all across the marketplace. In mobile networks, some of the most valuable of these applications are developing. They include tracking devices helping trucking firms manage shipping routes; OnStar radios that automatically call 9–1–1 when a car crash occurs (and airbags deploy); the embedded mobile phone connection in an Amazon Kindle, bringing seamless book downloads at the press of a button; sensors in vending machines that call a distributor when supplies of a product run low; or sensors in a person's body that ping the person's cellphone when his vital signs being monitored reach a critical level, notifying the patient and his doctors.

The machine-to-machine (M2M) applications space is just now forming, and already it is startling users, scientists, engineers, security experts and business logistics planners alike. The potential in just one field, that of mHealth, is by itself phenomenal. Consider the opportunities for monitoring the health of cellphone users, who live, work and sleep with a network-connected computer at their side; consider further the advances in medical science which will flow from the data gleaned from millions of such connected devices. Already, technology leaders such as Cisco are speaking of 2020 as a world of "50 billion things on the Internet."[42]

A defining characteristic of such smartphones is that they are designed to supply wireless broadband connections and to display multi-media content. Streaming audio and video applications have, hence, grown *pari passu* with smartphone penetration; as new applications and services come online to serve this burgeoning market, demand grows still further. This virtuous circle produces benefits for users and content vendors, but puts increasing pressure on mobile networks. In parallel, a defining characteristic of M2M devices is that—while some use very little bandwidth and others can consume a whole lot—they are fantastically voluminous in number. Their bounds are unknown. All of these devices put pressure on existing networks. As bandwidth consumption intensifies, operators have basically three ways to provide greater capacity:

- build more base stations—also known as cell-splitting;
- deploy more advanced technologies, upgrading base stations and handsets;
- use more spectrum.

Operators are doing the first two (by year-end 2010, carriers had constructed some 253,000 base stations, e.g.), but government largely controls the third. There is now great pressure for the FCC to put additional spectrum resources into the marketplace so as to give the mobile data tsunami the room it needs. In the National Broadband Plan, published in March 2010 by the Commission pursuant to a Congressional mandate, the FCC found that at least 300 MHz of new bandwidth would be needed by mobile carriers by 2015, and 500 MHz by 2020. It documented that, historically, spectrum allocations had been long and tedious, taking an average of 6 to 13 years to complete (an assessment that is exceeding generous in its calculations).

Where are the new airwaves for mobile to come from? The Commission recommended that up to 120 MHz of the 294 MHz digital TV band could be reallocated for mobile use, and identified other bands to bring the total new mobile allocations to the 300 MHz level. See Table 2.

Table 2. FCC Proposed Mobile Spectrum Allocations

Band	Key Actions and Timing	MHz Made Available
WCS	2010--Order	20
AWS 2/3	2010--Order 2011--Auction	60
D Block	2010—Order 2011--Auction	10
Mobile Satellite Services (MSS)	2010—L-Band and Big LEO Orders 2011—S-Band Order	90
Broadcast TV	2011—Order 2012/13—Auction 2015—Band transition/clearing	120
Total		300

Source: FCC, National Broadband Plan (March 2010), p. 84.

The plan is notable in its clear-eyed vision that more bandwidth is needed for mobile markets and that flexible-use licenses are the efficient way to supply such spectrum to the market. It is also blunt about the historical fact that regulators have a decided tendency to delay allocations, deterring exceptionally important economic growth. Its specificity in pinpointing the bands, and regulatory actions, needed to implement its proposal is also refreshingly transparent. Alas, it remains to be seen if such efforts will yet be sufficient to overcome the traditional sources of bureaucratic paralysis. Mobile markets will greatly benefit should they succeed.

Notes

1. George C. Calhoun, *Digital Cellular Radio* (1988).
2. Tom Farley, The Cell-Phone Revolution, *Invention & Technology* (Winter 2007); http://www.americanheritage.com/articles/magazine/it/2007/3/2007_3_8.shtml.
3. James Murray, Jr., *Wireless Nation* (2001), p. 24.
4. Jeffrey H. Rohlfs, Charles L. Jackson and Tracey E. Kelly, *Estimate of the Loss to the United States Caused by the Delay in Licensing Cellular Telecommunications* (Nov. 8, 1991); http://www.jacksons.net/EstimateofTheLossFromCellularDelay.pdf.

5. The FCC licenses both operators and devices. In the former case, a party is authorized to conduct certain types of wireless activities under terms and conditions established in the license. Specific bands are allocated to the license. TV stations, telecommunications satellites, and mobile telephone networks are authorized this way, with "licensed spectrum." In the latter case, the agency determines how radios will be permitted to operate, and then sets aside spectrum for use with such devices. This method is referenced as "unlicensed" or "license-exempt" spectrum.

6. Farley (2007).

7. Murray (2001), p. 50.

8. Ibid.

9. The standard required by cellular licenses was Advanced Mobile Phone Service, or AMPS. It was an analog technology that, given the long delays in licensing cellular, had already been eclipsed by emerging digital systems according to industry experts.

10. Licenses in rural markets went for a significant discount, given lower population density (and higher per-person capital costs). See Thomas W. Hazlett & Robert J. Michaels, The Cost of Rent Seeking: Evidence from the Cellular Telephone License Lotteries, 39 *Southern Economic Journal 425* (Jan. 1993).

11. National Telecommunications and Information Administration (Feb. 1991). The estimate for urban/suburban markets totaled about $80 billion. Reasonable rural market estimates increased the nationwide sum to approximately $100 billion.

12. This is true even for a monopoly wireless network. It is often asserted, incorrectly, that private licensees "hoard" airwaves so as to deter competition. But the profit-maximizing monopolist does not inefficiently set spectrum aside. Rather, he operates efficiently (minimizing costs) and then sets prices at a higher-than-competitive level.

13. Murray (2001), p. 105.

14. The new network, which incorporated an iDEN digital standard created by Motorola, would go on to serve 12 million subscribers and introduce a popular "push to talk" feature for "peer to peer" (phones transmit to each other rather than to the network) communications between a group of subscribers. It was sold to Sprint in 2005 for $35 billion.

15. Ellen Simon, *Sprint Acquires Nextel In $35 Billion Deal, CRN* (Dec. 15, 2004), http://www.crn.com/news/networking/55800041/sprint-acquires-nextel-in-35-billion-deal.htm;jsessionid=q1QzWlQIqI03u9d3HBuaDg**.ecappj02

16. The long license distribution period was caused by severe problems in extending subsidies, or "bidding credits," to small businesses and rural telephone carriers. These bidders were determined by the FCC to be "designated entities" (DEs) worthy of favorable rules for social policy reasons. In extending DEs long-term (10-year) financing at extremely favorable interest rates, the Commission set up a strategic game. DEs overbid for licenses, then declared bankruptcy, and then asked bankruptcy courts to reduce their liabilities (including monies owed the government for the licenses). The FCC fought these moves in federal court. In general, the DEs won, the FCC appealed, and the government lost again. For about a decade approximately 30 MHz of PCS spectrum was unavailable to the market, a loss to consumers of over $60 billion. See discussion below.

17. I served as Chief Economist of the Federal Communications Commission in 1991–92. I, too, gave the speech.

18. Source: CTIA. Note: In 2006–08, as text messaging (SMS) substitutes for voice minutes of use (MOU), MOU totals are calculated to include text messages at the "exchange rate" of 1 SMS = 1 MOU.

19. By 2005, all three major long-distance carriers—Sprint, MCI, and AT&T—had been rendered defunct. Sprint migrated its business to wireless; MCI and AT&T were absorbed

by local Bell companies (Verizon and SBC, respectively). (Upon buying AT&T, SBC took its the name.)

20. This takes the customer's total charges and divides by total usage.

21. For a detailed analysis, see Thomas W. Hazlett & Roberto E. Muñoz, A Welfare Analysis of Spectrum Allocation Policies, 40 *RAND Journal on Economics* 424 (Autumn 2009).

22. Verizon Wireless CDMA2000 1X Evolution, US, Mobilecomms-Technology.com; http://www.mobilecomms-technology.com/projects/verizon_3g/.

23. This was due to economies of scale. Adding a new network to an existing network does not necessarily increase spectrum per subscriber, but it does give the carrier more latitude in managing voice and data services across a larger bandwidth component.

24. Dan Richman, AT&T Wireless Sale to Cingular Finalized, *Seattle Post-Intelligencer* (Oct. 26, 2004); http://www.seattlepi.com/business/196839_attwireless26.html.

25. Cingular Wireless Plans Cellular Network Upgrade, *Dallas Morning News* (Dec. 1, 2004); http://www.redorbit.com/news/technology/107576/cingular_wireless_plans_cellular_network_upgrade/index.html.

26. "Cingular, which became the nation's largest mobile-phone provider last month by buying AT&T Wireless, had previously hinted it would start upgrading its network once the merger was complete and it had additional wireless spectrum." Ibid. "Cingular's recent acquisition of AT&T Wireless provided the company with the spectrum necessary to build the 3G networks." Cingular Press Release, Cingular to Deliver 3G Wireless Services, PR Newswire (Nov. 30, 2004); http://www.prnewswire.com/news-releases/cingular-to-deliver-3g-wireless-broadband-services-75599637.html.

27. Which Carriers Have 3G? *CNET News.com*; http://reviews.cnet.com/4520-3504_7-5664933-4.html.

28. Steve Rosenbush, The Logic Behind a Sprint-Nextel Deal, *Business Week* (Dec. 13, 2004); http://www.businessweek.com/technology/content/dec2004/tc20041213_3595_tc024.htm

29. Brian Dolan, T-Mobile USA's 3G: Better Late than Never? *Fierce Wireless* (Jan. 11, 2008); http://www.fiercewireless.com/story/t-mobile-usas-3g-better-late-never/2008-01-11.

30. Marguerite Reardon, T-Mobile Outlines Plans for 3G Network, *CNET News.com* (Oct. 6, 2006); http://news.cnet.com/T-Mobile-outlines-plans-for-3G-network/2100-1039_3-6123480.html.

31. Ibid.

32. Colin Giggs, T-Mobile USA to get rolling on 3G, *RCR Wireless* (May 2, 2008); http://www.rcrwireless.com/apps/pbcs.dll/article?AID=/20080502/FREE/686824713/1002/rss01.

33. Allie Winter, T-Mobile USA, Leap Move Ahead With AWS-1 Launches, *RCR Wireless*, (Oct. 16, 2008); http://www.rcrwireless.com/ARTICLE/20081016/WIRELESS/810139966/3g-buildouts-tied-to-spectrum-availability.

34. MetroPCS has actually pushed 4G services into the market using bandwidth acquired in the 2006 AWS auction. Mike Dano, MetroPCS to Skip 3G for LTE Rollout? *Fierce Wireless* (Aug. 3, 2010); http://www.fiercewireless.com/story/metropcs-skip-3g-lte-rollout/2010-08-03.

35. Federal Communications Commission, *Internet Access Services: Status as of Dec. 31, 2010* (Oct. 2011), Table 7, p. 24.

36. The FCC's transition to "Advanced Television" was triggered by requests, in 1985 and 1986, to use under-utilized UHF-TV channels for "land mobile" services. Reallocation might have happened early on, without any plan for transitioning television services to a digital TV standard, as the FCC ultimately decided to do. Reallocation of UHF then queued up behind the transition. The FCC proceeding to achieve that began in 1987. The last full-

power analog TV stations went off the air in June 2009. On the origins of the "digital TV transition," see Joel Brinkley, *Defining Vision* (1997).

37. Mark Sullivan, Verizon Launches LTE "4G" Service in 39 Cities December 5, *PC World* (Dec. 5, 2010); http://www.pcworld.com/article/212121/verizon_launches_4g_lte_service_in_39_cities_december_5.html.

38. Mike Dano, MetroPCS to Skip 3G with LTE Rollout? *Fierce Wireless* (Aug. 3, 2010); http://www.fiercewireless.com/story/metropcs-skip-3g-lte-rollout/2010–08–03.

39. Ibid.

40. Of course, in parts of the country where there is little population and, hence, low demand, that margin may be much closer.

41. The classic critique of this "command and control" system was delivered by Ronald H. Coase, The Federal Communications Commission, 2 *Journal of Law & Economics* 1 (1959). a more recent analysis, referencing traditional spectrum allocation as analogous to "Gosplan," see Gerald R. Faulhaber & David J. Farber, Spectrum Management: Property Rights, Markets and the Commons, AEI-Brookings Joint Center for Regulatory Studies Working Paper No. 02–12 (Dec. 2002).

42. Arik Hesseldahl, Cisco Reminds Us Once Again How Big the Internet Is, and How Big It's Getting, *All Things Digital* (July 14, 2011); http://allthingsd.com/20110714/cisco-reminds-us-once-again-how-big-the-internet-is-and-how-big-its-getting/

6

Not TV, Not the Web

Mobile Video Between Openness and Control

AYMAR JEAN CHRISTIAN

> Without exception, the brave new technologies of the twentieth century—free use of which was originally encouraged, for the sake of further invention and individual expression—eventually evolved into privately controlled industrial behemoths, the "old media" giants of the twenty-first, through which the flow and nature of content would be strictly controlled for reasons of commerce.[1]
>
> —TIM WU

At the end of the 2000s a rhetoric about mobile emerged, somewhere between openness and control, possibilities and restrictions. After years of legal tussles, the Federal Communication Commission voted on December 2010 on rules for net neutrality, proposing light regulation of wired broadband to "to preserve the open Internet" yet carving out exception for wireless services.[2] Months before the FCC ruling, Chris Anderson, already a prominent theorist of new media,[3] had penned a much debated op-ed proclaiming "the web is dead," lost to the "semi-closed" world of mostly-mobile apps: "Over the past few years, one of the most important shifts in the digital world has been the move from the wide-open Web to semi-closed platforms that use the Internet for transport but not the browser for display."[4] At the heart of Anderson's critique was the idea of mobile, wireless applications, which stood in opposition to the exploratory and serendipitous possibilities of the Internet.

The lesson was clear: the web was open and the new frontier, mobile, was not—or not completely.[5] In the wake of debates over net neutrality and the changing nature of the Internet, mobile technologies have been marketed to the public as "different" from other forms of distribution, a separate sphere, at once a space of openness and possibility and one controlled by major corporations looking to restrict its potential richness and diversity. At the heart of these debates over what mobile meant was video, whose large demands on data raised the stakes for all the players invested in the medium. Mobile video lacked a clear ideology: not "closed" like TV, whose content providers and distributors were largely established, and not "open" like the web, whose participants still enjoyed relatively low barriers to entry either from platforms like YouTube or dozens of alternative start-ups from Blip TV to Funny or Die.

Yet this sweeping rhetoric, embodied in many ways by scholars like Tim Wu, fails to capture the intricacies of how the mobile market works. What an examination into the mobile market will show is how the medium remains at once highly volatile, deeply complex and somewhat confusing for video distributors, who must navigate numerous layers of the industry to reach users. This complexity suggests mobile is a space of compromise, possibility and challenges that occasionally requires players to choose between open and closed options.

This chapter focuses on the efforts of three distributors of independent web video—Vimeo,[6] My Damn Channel, and Q3030 Networks—alongside larger video sites—YouTube, Hulu and Crackle—to show how navigating the mobile market involves negotiating industrial and technological considerations, sometimes open and closed, sometimes neither. I outline what these companies wanted from mobile and how they conceptualized their needs in the months leading up to and directly following the government's first official statement on net neutrality and its exception for wireless services. From their perspective, the realities of the mobile video market demonstrate how new media arise in fractured markets. This chapter analyzes a sector of the mobile video market in a specific, narrow period of time. By focusing on "mobile video" it leaves out efforts to broadcast television signals to mobile phones, instead looking at using cellular and potentially broadband networks to deliver web-like signals to phones.[7] In the end, the mobile device itself holds no inherent meaning or politics outside its market and government players, all of whom are still working out how to deliver mobile content.

The Deep Complexity of Mobile Video

The mobile media market is a complex industrial and technological system with different barriers to entry than the web. Controlled much more tightly by network providers, and with less bandwidth than either TV or web, the history of developing video content for mobile devices has been as stunted as the prior developments of radio, television and web video.

Before delving into how independent and corporate video providers navigate the mobile market, it is important to understand generally how that market works. At each level of the mobile market are competing claims for whether the medium is "open"—able to incorporate diversity and citizen participation—or "controlled"— at the ultimate whims of corporations. But the dichotomy is limiting. History shows periods of new media engender efforts to control the new medium, and in the wake of federal regulation, the mobile video market is in a tussle for whom will win, not necessarily what is best for the consumer or content providers.

From roughly 2009 to 2011, much of the buzz in the mobile market concerned the rapid adaptation of smartphones, devices that could handle web services with large enough screens for viewing video.[8] Even as delivering video became simpler, the realities of the mobile market remained in flux on several levels: the phone market (hardware), the network (cellular, wireless, broadcast), the operating system (from Google's Android to Apple's iOS), software and compatibility (DRM, Flash), and finally the application layer, in which companies must determine how users will access services.

Before the rise of Apple's iOS devices and Android-enabled smartphones, the mobile video market relied heavily of the capabilities of individual phones and devices, whose operating systems were dictated and negotiated by cellular network providers and phone, tablet and netbook manufacturers.[9] Networks like Verizon and Sprint dictated what kinds of videos users could watch largely based on deals struck with content providers like Viacom:[10] "The various cellular providers have become new gate-keepers, and their decisions about which handsets and services to support are a fundamental parameter for the development of mobile media."[11] The advantage of operating systems like Android and iOS, which grew more popular through 2010, was how they streamlined the development process for content providers.

Yet before content reaches the phone, it must be delivered through networks carrying data. The heavy bandwidth demands of video in many ways stunted adoption of wireless video and raised the stakes for players in the mobile market, each of whom claimed their method for delivered wireless video was in the public interest. Cellular network providers advocated for "mobile broadband" which could transfer the "open" aspects of the Internet to mobile: "Wireless Internet access provides broadband to the person at anytime and anywhere. The ability to access the Internet, regardless of your location, is the great equalizer."[12] Such pronouncements were in rhetorical opposition to traditional broadcasters, who wanted to keep enough of their spectrum to allow local broadcasters (and other networks) "to jump into the future with mobility": "Mobile DTV is giving you the ability to carry broadcast TV wherever you go."[13] As the FCC was deciding not to regulate wireless, it was also negotiating a plan to transfer spectrum to network providers, who wanted to charge to faster mobile speeds as demand for mobile video increased.[14] For video providers, this was, on its face, a good thing, as it could raise their viewer-bases. But it also would give more power to cellular network providers, who could turn on companies like Netflix and Vimeo.[15] Which system is more open remains an unanswered question, with industry groups from mobile broadband-friendly CTIA to the broadcaster-friendly Open Mobile Video Coalition advocating for two different paradigms. Nonetheless, mobile broadband, supported in part by the rhetoric of the open web, has the upper hand.[16]

Largely isolated from debates about the network, video content developers grappled with competing claims to "open" and "closed" operating systems: iOS and Android. Apple's successful entrée into the mobile market with the iPhone and then iPad gave it an early lead and many providers like Netflix created content for its devices first. Apple's dominance, its availability only on Apple devices and its taking a cut of application sales led to criticisms it was a "closed platform."[17] Allegations surfaced that Apple slowed down apps built to bypass its own Store.[18] Android, in the meantime, was open to developers and could be placed on any phone and supported video with open codecs.[19] But evidence suggested Google used Android's compatibility standards inconsistently, sometimes forcing developers out of contracts[20] and allowing carriers like Sprint and Verizon to load "propriety ecosystems" on Android-enabled phones.[21] These "ecosystems" were video stores that competed with the kind of smaller video distributors this chapter will discuss::[22]

> Is Android more 'open' than iOS is? Yes. But the way Google has been throwing around the word is in absolute terms. It has been "open" (them) versus "closed" (Apple). That's simply not true. And in that context, being "more" open is like being "kind of" pregnant.[23]

Video providers looking to reach users on mobile phones see past these debates over openness and talk in nuanced ways about the subtle difficulties, and pros and cons, of developing for various operating systems.

Delivering video to mobile devices requires an intimate knowledge of debates over compatibility standards, playback software and codecs. Within this tech-driven world of content delivery, the largest debate lies between HTML5, a new, presumably more open technology, and Flash, the standard playback mechanism for web video. At the heart of this debate is, again, video, and distributors like Vimeo and My Damn Channel must decide to what degree they will engage the debate. The lines of division are not entirely clear: Apple, the assumed "closed" system, supports "open" HTML5, claiming "…Flash as a closed, proprietary technology that gobbled up too much of the CPU's resources and compromised battery life."[24] However, some claim Apple supports HTML5 because supporting Flash would cannibalize its pay-based iTunes market before the company creates a cloud video service.[25] Meanwhile Flash's dominance, which it has gained for both its quality playback and support for Digital Rights Management, a key component for corporate content producers, has eased the tech development process for independent distributors on the web. The mobile space's fast adoption of HTML5 through 2010 created compatibility woes for numerous video creators accustomed to Flash online.[26]

Tied in with these debates is the ongoing contest between applications (apps) or browser-enabled mobile surfing. Each mode of delivering has, it will be shown, its claims to openness and user-friendliness, a still uneasy terrain for video distributors.

Vimeo: Navigating Mobile for Independent Filmmakers

Vimeo's long road to delivering mobile video illuminates the set of challenges facing distributors in periods of media transition. "Delivering video is one of the hardest things you can do," Andrew Pile, Vimeo's vice president of product and technology said in an interview.[27] "There's so many gadgets, and there's so many browsers and different limitations, and now with all these devices out there, there's less power, less compatibility." As a video aggregator Vimeo is a smaller player than its larger counterpart, YouTube, but over its five-year existence has carved out a strong and sizable niche as a purveyor of higher quality independent video, an image the site has cultivated.[28] As the distributor of choice for thousands of independent filmmakers, Vimeo felt compelled to enter the mobile space. "It's just becoming standard," Pile said.

The debates around the openness of the mobile medium were less relevant to Vimeo's concerns as was grappling with multiple levels of fragmentation: the plethora of devices, the range of playback software, and the decision of whether to support web-enabled mobile browsers or self-contained applications for smartphones and tablets. Navigating this terrain, Vimeo came up against the open or closed capabilities of various options, but creating a mobile experience for video creators meant finding flexible solutions that worked on competing platforms.

Vimeo started its push to get on mobile devices in late 2009 and early 2010. From the beginning the company had to choose whether it would develop a better browser experience or a separate app. It opted for both, but getting the browser—

the web experience—to work was a priority. Much of the site's traffic came from video embedded outside of the home site of Vimeo.com, dispersed across the open web, and it needed to support that growing ecosystem. "That content was getting shared on different web pages and now it's getting shared on different devices with different capabilities," Pile said.

But creating a mobile player capable of matching the openness of the web proved difficult. The mobile browser needed a video player able to deliver video on a range of devices from iPhones and Android phones, to tablets like the iPad and less video-friendly devices like the BlackBerry, all of which already had built-in YouTube-capability because of that site's dominance in the market. The absence of a playback software monopoly in mobile—Flash dominated the web at the time—meant Vimeo needed to convert its Flash-based web player to a "universal player" that fit within various systems and served different video formats to different devices with agility. "You need to be able to support multiple formats, multiple players, multiple devices, multiple platforms," Pile said. "What used to be one Flash player, there's many different combinations of things it could be. It could be an HTML5 touch player, or it could be a Flash mobile player. So now we have maybe 6 or 8 different possible combinations of what we're serving." Flash, a propriety DRM-friendly technology, then, was a major support for the presumed openness of the web; meanwhile, the mobile space, utilizing numerous technologies, presented a far greater challenge. HTML5 worked better across devices—hence its brand as more open—but had fewer features and less efficiency; Flash, whose mobile performance was also suspect, delivered a better experience on the web.[29] The solution was Vimeo's current embed code using <iframe> technology; for this, video on a mobile site works by actually delivering a webpage, not the video itself, showing how crucial the web remains to mobile technology. This flexibility is workable but temporary solution for the still fragmented market.

For Pile, one of the most important axioms of managing video technology is that users "expect it to just all work," meaning distributors need to invest the necessary labor to streamline the chaotic technological and industrial landscape of mobile media. But perfection is not achievable, and some "guesswork" is necessary—for instance, when its embeddable player has to distinguish between an iPhone 3 and an iPhone 4, which have different capabilities for handling high quality videos. To create ease of use, the company has multiple players for its videos, each with different capabilities: a mobile-style player with limited additional functions like "sharing" and "liking;" a tablet player with a more desktop-style look; and the "native player," in which Vimeo videos load on whatever that device's standard player is (Quicktime, for example, on the iPhone), though sometimes the site can make the native player look like a Vimeo video. "We do a lot. We spend a lot of time tweeking…We probably have a dozen phones sitting in desk drawers," Pile said. Even still, some devices are simply not video-friendly: "Being on BlackBerry, it's definitely a hostile Internet out there."

With all of these factors going into building video for mobile browsers, Vimeo originally thought creating a mobile app would be simple. But the separate user expectations for individualized apps made the process rocky. Unlike on the web, users expected "closed" apps to do everything that was possible on the web and more. After delaying an app release for a number of years, when the site first ventured into the space, it had conceived of creating a program to help users shoot and edit video but kept adding features: "What was going to be a 2-month project turned into an 8 month project," Pile said. Even after completion, Vimeo ran into some trouble when its app did not allow search. "We had literally hundreds of features in there: you could edit videos for free, upload them for free, share them, all this great stuff, but, oh, it didn't have this one feature.…It was a much more aggressive world than we were expecting." As intricate as the web-mobile space was, Vimeo realized users were forgiving of browser-based mobile experiences but not apps: "You take the whole Internet and squeeze it into a phone and it doesn't quite work right, people say 'oh, it's the phone's fault.'" But for apps, whose software is harder to master, "people really expect it to be nice, because it was built for that experience. So if it's off, people seem to be more attuned to that."

In the end, Vimeo's experiences building a video platform that works for mobile suggests a much more fragmented and complicated story, the hidden ruptures in convergence culture.[30] The company nevertheless envisions a future with a more open and flexible video market: "There's definitely going to be this platform-agnostic stuff going on. The expectation will definitely be 'this video should *work* everywhere, this video should *be* everywhere, I should be able to get to this video everywhere.'" A truly flexible mobile video experience, capable of moving through the various layers of the industry, remains a bit of fantasy for video distributors in this specific historical moment, and one that requires a lot of labor.

My Damn Channel and Q3030: Independents Hoping to Break Through in a Fragmented Market

For most companies, mobile video is just one part of a broader move toward convergence: the ability to access content across TV, computer and mobile screens. What this means for independent distributors is they have to work harder to optimize their content with a widening array of platforms, from those that connect the web to TV—Apple TV, Roku, Boxee—to those that optimize the web for mobile screens—

primarily iOS and Android. Developing for television has its own constraints, and while mobile is less complicated, it brings with it a range of complicated decisions, including whether or not to create an app, how to optimize the user experience or develop for new devices. For these distributors, issues of openness do arise—especially when it comes to iOS and Android—but developers spend more time navigating mobile's fragmented universe.

Q3030 Networks is an interesting anomaly. The independent network debuted in 2011, focusing on distribution everywhere *but* the web. While the network has a dedicated web page, as is standard for most video distributors, it is making a big push for web-to-TV set-top boxes and a range of mobile platforms. "For us it's about having a broader mind. We are calling ourselves a connected device network. Whatever, whenever people can watch video on, we're trying to be there," Marq Sears, the company's founder and CEO said in an interview.[31] This platform-agnosticism, however, is not easy. The network has a team of programmers working constantly to develop applications for a variety of devices in the U.S. and U.K. "We have to program for every platform....There is a barrier to getting involved....That's why you don't see people running towards it."

The app market—the market Chris Anderson feels kills the open web—creates its own challenges for purveyors of mobile video. Users demand more from mobile applications, as an enclosed, if not closed, entity. For Q3030, that meant as much time was spent on the aesthetics of their apps as their technology: "The most challenging part for us is the UI [user interface], how it looks and feels, and then make sure it's super-compelling." Q3030 is trying to avoid the stumbles of companies like Vimeo and take cues from distributors like HBO, focusing on aesthetics and ease-of-use. This includes "search and discovery," something natural to the web but that raises questions on mobile devices: "When you initially access it, how intuitive is the app? How do you get to shows? What shows up? And how clean is the design?"

For a new video company the app world is far from closed, or semi-closed. Indeed Sears' rhetoric about the potential for mobile media echoes earlier ideas about the open web—at least the pre-Google, Facebook and net neutrality web—emphasizing lower barriers to entry, the possibility of serendipity and new audiences for independent content:

> There's a search and discovery that's going to happen just because you're in the app store. You're going to get people who climb aboard just because they found you in there. They walked through the market, they were just searching through and see, "oh, media network. Click!" So you get all that. You get search and discovery of viewers.[32]

Sears' invocation of an open market, mirroring arguments made by the digital optimists of *The Cluetrain Manifesto* and writers like Anderson, suggests the app world as space of possibility for video distributors. Yet for Sears the app world is also about exclusion, or exclusivity. For Q3030, the app market was a way to reach app users—people with smartphones and devices—who are more savvy and therefore more marketable than regular web users.[33] Moreover, the "walled garden" approach to app markets, particularly on the iPhone, appealed to Sears: "By being inside the app store it creates a certain kind of validity for your brand….Apple has accepted you, or Android has accepted you. It's kind of like an official seal….It's a subconscious purchase-play on the consumer." The app market is open, then, after first being closed. Even still there are barriers to entry into app markets, particularly for iOS. Sears knows it could take months before Apple approves an app. "Android's a little easier because of the openness of the platform for development. There's no hurdles really on the Android side." The meaning of the app world, therefore, is still being negotiated.

My Damn Channel, a five-year-old independent comedy network distributing web series from celebrities and major brands, started its mobile push by focusing not on apps but rather, like Vimeo, on optimizing the web experience on mobile phones and tablets. The site, however, didn't come to this decision until after Android and iOS-enabled phones hit the market. Before, each carrier had—and still has—different platforms for delivering video, making it difficult for an independent like My Damn Channel (corporate-owned sites like Crackle and YouTube were able to negotiate deals). Now, according to Warren Chao, My Damn Channel's chief operating officer, "you can develop for two platforms and capture most of the market."[34]

The transition to mobile, however, was not seamless. My Damn Channel settled a temporary solution for iOS devices, utilizing YouTube's player to allow users to play videos on its site on mobile devices.[35] Here, YouTube's early lead in the mobile space—its capability on most devices because of its ubiquity—worked in the comedy network's favor. For Android devices, the operating system's "openness," its support for Flash, allowed My Damn Channel to work relatively well on those devices. But neither of these two solutions were permanent, neither a perfect fit for both the closed and open operating systems. My Damn Channel's Flash player was less than efficient on Android and on iOS its YouTube player bypasses the site's own ad-delivery system.[36] Chao appreciated Google's openness to Flash—"They're not as closed a platform…yet"—but Flash's mixed performance overall and Apple's lack of support for the software has pushed them to rethink their tech decision and move away from Flash.

Part of that decision, a way to bypass the weaknesses of both operating systems, is to create applications for smartphones and tablets like Q3030. Jumping into the app market, as this essay shows, is not easy. "If you're going to launch an app you have to make it different and tap more functionality than you would for just a simple version of your website that's enabled for a browser," Chao said. For My Damn Channel, apps are "different" from the web, their priority. Mobile browsers, more so than Internet browsers, are enmeshed in battles over presumably open and closed software like Flash, DRM and HTML5. Apps offer another option, one that many other companies have taken, but put new burdens on websites, particularly small independent ones.

Corporate Video: Leveraging Size to Dominate a New Space

For the relatively small players above, delivering mobile video requires trial and error, compromises and tough calls. For corporate video distributors, the story is different. Mobile video offers the potential for additional revenue streams—the web is free, mobile is undefined—and companies have historically used the medium to strike deals with telecommunications oligarchs. Yet many of these distributors have used the growing market for smartphones to focus on free, ad-supported websites and apps, though to what degree those offerings are free or open remains up-for-debate. The mobile video market for these players remains similarly inchoate but rife with possibility for gaining market share against competitors.

If mobile video is between openness and control, YouTube's history in the space indicates it might value openness. YouTube was the earliest and biggest video player in mobile, and it continues to grab the highest traffic in the market. Its large library of videos and popularity as a purveyor of viral (or spreadable) video has caused it to move away from the kinds of deals it had historically made with cell providers, which allowed access to YouTube only after subscribers paid fees,[37] to utilizing special apps on smartphones like the BlackBerry and iPhone, to developing a mobile website that carried the Internet desktop experience across platforms.[38] The rhetoric of openness here—from a "device-centric world to a use-centric world" like the web—supports the company's larger desire to become more than a destination, to dominate the video market as an "operating system."[39] A major part of this transition was Google's rapid transition to the presumably more open and flexible HTML5, yet the company's support of the playback standard, which it hopes will allow users greater access to YouTube content across devices, has forced others to adopt it as well.[40]

For YouTube, then, the mobile space is a way for the company to translate and solidify its web dominance on new devices, even as it justifies its ambitions as opening up the medium. For other major video players including Hulu, Netflix and

Crackle, already accustomed to ranking behind YouTube, the mobile space's immaturity offers them the chance to get users accustomed to paying for content. Hulu made its entrée into mobile distribution through its Plus service and has resisted attempts to allow playback on web-enabled mobile browsers that "challenge the need for apps:" "[Hulu] has adopted a strategy of denying consumers free access to its content through web-connected televisions and new mobile devices."[41] In this case, Hulu, whose corporate video is arguably more desirable than the independent- and amateur-friendly YouTube, had been too open on the web and used mobile to develop paying users.[42]

Crackle adopted a similar strategy early on, negotiating deals with Verizon's VCast, with Sprint through MobiTV's subscription-based platform, with the now-deceased FLO TV and though its own pay-per-view subscription service.[43] Yet with the advent of smartphones, Crackle saw an opportunity to create ad-supported applications that competed with the subscription-based offerings of Netflix and Hulu and worked across platforms. According to Crackle head Eric Berger: "What the consumer is getting used to through subscription services (such as Netflix and Hulu) are now available through this ad-supported service where you can log into Crackle, make a queue on your iPad, go finish it on your PC or go finish it on a Google TV or on a PlayStation and manage that queue."[44] The "free" option works for Crackle, in part because the ads it delivers cannot be fast-forwarded—and unlike on the web, users have less flexibility to avoid ads by shifting browsers or applications during playback.[45]

Mobile Video in Search of Openness

Around the time of the FCC's net neutrality decision, a very convincing rhetoric about the history of media started to emerge, promulgated most visibly by Tim Wu. For Wu, the history of the 20th century was one of open new media transitioning to closed platforms owned by corporations. Does the mobile media market adhere to this teleology? Yes and no. Ignoring the fact that access to mobile technology has always been costly to consumers, this essay has shown how developers, large and small, cannot completely rely on variously open and closed technologies and business decisions to deliver video to users. Mobile is unlike radio—which allowed for relatively open access to distribution in its early years—television—which had a small space for Hollywood independents—and film—whose early years saw numerous upheavals over production and distribution.[46] The mobile frontier attracted entrepreneurs, large and small, mainly because it was growing and because customers seemed willing to pay for content—though it turns out only in limited circumstances.[47]

For video producers and distributors, mobile was already hardwired to not be amendable to independent creation: the operating systems and software were created by companies and standards were already in place, the network was managed by an oligopoly, and there was limited bandwidth. Moreover, users were already used to companies managing the network (as opposed to the web, which grew out of the government and universities). Meanwhile companies, who often have little ideological commitment to notions of openness, are using mobile as a testing ground for their ability to achieve financial stability in an era of convergence. Mobile media still have the potential to be open for video, but such an ideal must be actively pursued both within and outside of the industry.

Notes

1. Wu, T. *The Master Switch: The Rise and Fall of Information Empires* (New York, NY: Random House, 2010), 6.
2. Kim, Ryan. "FCC's New Net Neutrality Rules to Regulate Wireless Lightly," *GigaOM*, December 1, 2010, http://gigaom.com/2010/12/01/fccs-new-net-neutrality-rules-to-regulate-wireless-lightly. For a thorough explanation on the history of net neutrality and the wireless exception, see Holt, Jennifer. "Platforms and Pipelines in Transition: Anatomy of a Policy Crisis." Presentation at Media in Transition 7, Unstable platforms: the promise and peril of transition, Boston, MA, March 13–15, 2011.
3. Anderson, Chris. *The Long Tail: Why the Future of Business Is Selling Less of More* (New York, NY: Hyperion, 2008); and Anderson, Chris. *Free: The Future of a Radical Price* (New York, NY: Hyperion, 2009).
4. Anderson, Chris. "The Web Is Dead. Long Live the Internet," *Wired*, September 1, 2010.
5. The rhetoric about the death of the "open" web was picked up in scholarship too. See Turow, Joseph. *The Daily You: How the New Advertising Industry is Defining Your Identity and Your Worth* (Yale University Press, 2012); Vaidhyanathan, Siva. *The Googlization of Everything: and Why We Should Worry* (Los Angeles: University of California Press, 2011).
6. Vimeo is owned by IAC/InterActiv Corp. but does not allow commercial videos, in-stream advertising and is largely supported by paying users.
7. Arceneaux, Noah. "'All You'll Need Is a Mobile Couch': The History of Mobile Television in the United States," in *The Long History of New Media*, ed. David W. Park, Nicholas W. Jankowski, and Steve Jones (New York: Peter Lang, 2011), 21–36.
8. MacManus, Richard. "Top 5 Web Trends of 2009: Mobile Web & Augmented Reality," *ReadWriteWeb*, September 10, 2009, http://www.readwriteweb.com/archives/top_5_web_trends_of_2009_mobile_web_augmented_reality.php.
9. Some argue it still needs to open up: "It's time for content providers to embrace the web and set our mobile media free because limiting it to certain platforms or handsets is likely also limiting sales: the more freedom consumers have to move content around to the time, place and device of their choosing, the more apt they are to rent or buy that video." Tofel, Kevin C. "Mobile Media Needs to Be Set Free," *GigaOM*, April 12, 2011, http://gigaom.com/mobile/mobile-media-needs-to-be-set-free/.

10. Video distributors would announce special deals with cellular providers, who then often locked content behind expensive data plans. See, as one example, one of the earliest deals, between YouTube and Verzion: Noguchi, Yuki. "Hello, Cellphone? YouTube Calling," *Washington Post*, November 28, 2006, D1.

11. Arceneaux.

12. Largent, Steve. "CTIA-The Wireless Assocation Responds to NAB on Spectrum," *CTIA-The Wireless Association Blog*, January 31, 2011, http://blog.ctia.org/2011/01/31/ctia-the-wireless-association-responds-to-nab-on-spectrum.

13. Personal interview with Dave Arland, April 19, 2011 spokesperson for the Open Mobile Video Coalition.

14. Kim, Ryan. "FCC Moves to Free Up TV Airwaves for Internet Use," *GigaOM*, November 30, 2010, http://gigaom.com/2010/11/30/fcc-moves-to-free-up-tv-airwaves-for-internet-use. For information on demand for mobile bandwidth, see Nielsen. "More Americans Watching Mobile Video," *Nielsen Blog*, December 8, 2010, http://blog.nielsen.com/nielsen-wire/online_mobile/americans-watch-more-mobile-video-now-than-ever/; and Lunden, Ingrid. "Video Is 40% of All Mobile Traffic, but Most of Us Don't Watch It," *paidContent*, January 18, 2011, http://paidcontent.org/article/419-video-is-40-of-all-mobile-traffic-but-most-of-us-dont-watch-it/.

15. Higginbotham, Stacey. "Decision Time: Does the Nation Need TV or Mobile Broadband?," *GigaOM*, October 28, 2009, http://gigaom.com/2009/10/28/decision-time-does-the-nation-need-tv-or-mobile-broadband.

16. Higginbotham, Stacey. "Video Haunts and Shapes the Spectrum Debate," *GigaOM*, May 20, 2011, http://gigaom.com/broadband/video-haunts-and-shapes-the-spectrum-debate.

17. Evans, Jonny. "Why 'Open' Android May Lose the Apple iOS Wars," *Computerworld*, August 24, 2010, http://m.blogs.computerworld.com/16816/android_army_plans_ipad_overthrow.

18. Watters, Audrey. "Report Finds Performance of Web Apps Throttled on iOS Devices," *ReadWriteWeb*, March 15, 2011, http://www.readwriteweb.com/archives/report_finds_performance_of_web_apps_throttled_on.php.

19. See: Melanson, Mike. "Google Says It's Open or Not at All for Video on Chrome," *Read Write Web*, January 11, 2011, http://www.readwriteweb.com/archives/google_says_its_open _or_not_at_all_for_video_on_ch.php; and Arthur, Charles. "Google's WebM v H.264: Who Wins and Loses in the Video Codec Wars?," *The Guardian*, January 17, 2011, http://www.guardian.co.uk/technology/blog/2011/jan/17/google-webm-vp8-video-htm15-h264-winners-losers.

20. Rowinski, Dan. "Is Android Really Open?," *ReadWriteWeb*, May 6, 2011, http://www.readwriteweb.com/archives/is_google_stifling_innovation_in_battle_with_skyho.php

21. Fehrenbacher, Katie. "Verizon's VCAST, the no-fun network," *GigaOM*, February 15, 2007, http://gigaom.com/2007/02/15/vcast-warning.

22. Siegler, M. G. "Android Is as Open as the Clenched Fist I'd Like to Punch the Carriers With," *Techcrunch*, September 9, 2010, http://techcrunch.com/2010/09/09/android-open.

23. Siegler, M. G. "'Open,'" *Techcrunch*, March 26, 2011, http://techcrunch.com/2011/03/26/open.

24. Idelson, Karen. "Flash vs. HTML5: Don't Hate the Player," *Variety*, November 24, 2010, http://www.variety.com/article/VR1118027963.

25. Ong, Josh. "Google to preempt Apple's iCloud with New Music Service on Tuesday." Apple Insider, May 9, 2011. http://www.appleinsider.com/articles/11/05/09/google_to_preempt_apples_icloud_with_new_music_service_on_tuesday.html.

26. Gonsalves, Antone. "Vimeo Announces Flash, HTML 5 Hybrid Video Player," *InformationWeek,* August 17, 2010, http://www.informationweek.com/news/software/soa_webservices/226700407.

27. Personal interview with Andrew Pile, May 5, 2011.

28. Graham, Jefferson. "Vimeo Builds a Sense of Unity," *USA Today,* January 26, 2010, http://www.usatoday.com/MONEY/usaedition/2010–01–27-vime027_ST_NU.htm.

29. Perez, Sarah. "Does HTML5 Really Beat Flash? The Surprising Results of New Tests," *ReadWriteWeb,* March 10, 2010, http://www.readwriteweb.com/archives/does_html5_really_beat_flash_surprising_results_of_new_tests.php.

30. Vimeo is also on TV through set-top boxes, but Pile notes how navigating the even more propriety space of web-to-TV remains a painstaking process.

31. Personal interview with Marq Sears, May 9, 2011.

32. This rhetoric was quite pervasive throughout the 2000s. See Levine, Rick, Locke, Christopher, Searls, Doc and David Weinberger. *The Cluetrain Manifesto: The End of Business as Usual* (New York, NY: Basic Books, 2001); Shirky, Clay. *Here Comes Everybody: The Power of Organizing without Organizations* (New York, NY: Penguin Books, 2008); Wu, 2010.

33. Chris Anderson's concern about the app market, how it privatizes and streamlines the chaos of the web, may then prove justified. The app market as a semi-closed, or at least controlled, space, fits within larger discourses of mobile media as "privatized," in a number of ways. See: Groening, Stephen. "From 'a Box in the Theater of the World' to 'the World as Your Living Room': Cellular Phones, Television and Mobile Privatization," *New Media & Society* 12 (2010): 1331.

34. Personal interview with Warren Chao, May 3, 2011.

35. My Damn Channel has been distributing on YouTube for years and built a considerable audience on the site.

36. YouTube takes a cut of the profits. My Damn Channel has its own custom, Flash-based player.

37. Noguchi, 2006.

38. Bilton, Nick. "Google Makes the iPhone YouTube App Obsolete," *New York Times: Bits Blog,* July 7, 2010, http://bits.blogs.nytimes.com/2010/07/07/google-makes-the-iphone-youtube-app-obsolete.

39. Hachman, Mark. "YouTube Reworks Mobile Site for Speed, HTML5," *PC Magazine,* July 7, 2010, http://www.pcmag.com/article2/0,2817,2366159,00.asp.

40. Halliday, Josh. "HTML5 Version of YouTube Launches for Mobiles," *The Guardian,* August 25, 2010, http://www.guardian.co.uk/technology/pda/2010/aug/24/html5-youtube-mobile.

41. Worden, Nat. "Hulu Blocks Access to Its Content on RIM's PlayBook Tablet," *Dow Jones Newswires,* April 21, 2011, http://online.wsj.com/article/SB10001424052748703983704576277151165572370.html.

42. Learmonth, Michael. "Hulu's a Towering Success—Just about Every Way but Financially; Web-Video Future Hinges On It Working Out Pay Model," *Advertising Age,* March 29, 2010, 0001.

43. MobiTV Inc. "MobiTV, Inc. and Crackle Announce Major Viewership Milestone; 1 Hour+ Average Viewing Time Shows Mobile Users Crave Movies," *PR Newswire*, October 25, 2010. Domanick, Andrea. "SPT's Eric Berger: Digital Networks Are No 'Add-On'; Web Is at Core of Multi-platform Crackle's Content Creation and Distribution," *Broadcasting & Cable*, August 23, 2010; "Sony Pictures Television; Crackle Extends Its Premium Movie Service to All FLO TV Direct-to-Consumer Offerings," *Entertainment Newsweekly*, April 9, 2010; and Weprin, Alex. "Eric Berger; Senior VP of Digital Networks, Sony Pictures Television," *Broadcasting & Cable*, June 8, 2009, 11.

44. Snider, Mike. "Sony's New Crackle App for iPad and iPhone Lets You Take 'Seinfeld' with You," *USA Today*, April 18, 2011, http://content.usatoday.com/communities/technologylive/post/2011/04/sonys-new-crackle-app-for-ipad-and-iphone-lets-you-take-seinfeld-with-you/1.

45. Bond, Paul. "Sony Releases Crackle App for Apple Devices," *Hollywood Reporter*, April 18, 2011, http://www.hollywoodreporter.com/news/sony-releases-crackle-app-apple-179499.

46. For radio, see Douglas, Susan. *Inventing American Broadcasting, 1899–1922* (Baltimore: Johns Hopkins University Press, 1987); Hilmes, Michele. *Radio Voices: American Broadcasting, 1922–1952* (Minneapolis: University of Minnesota, 1997); Hilmes, Michele. *Hollywood & Broadcasting: From Radio to Cable* (Chicago: University of Illinois Press, 1999). For television, see Mann, Denise. *Hollywood Independents: The Postwar Talent Takeover* (Minneapolis: University of Minnesota Press 2008). For film see Wu, 2010.

47. Scott, Jeremy. "75% of Consumers Are Willing to Pay for Web Video Content," *Reel SEO*. Last accessed July 1, 2011. http://www.reelseo.com/75-consumers-pay-online-video-content.

7

Reading After the Phone

E-readers and Mobile Media

GERARD GOGGIN & CAROLINE HAMILTON

> It is hard to see how, and in what ways, the "new" technologies *are* truly new differ-
> ent, and *do* represent a radical break with the past, and in a sense just more of the same.
> —DAVID M. LEVY, *SCROLLING FORWARD*[1]

Introduction

For some centuries, reading has been associated with portable, personal media technology. Many scholars have noted and praised the mobility of reading brought about
by the emergence of the book and the advent of early modern print culture.
However, the links between reading (and writing), mobility, and media stretch back
further still, as observed by German media theorist Friedrich Kittler:

> Writing...served not only as a storage medium for everyday spoken language, but
> also...as a very slow broadcast medium after the practice of inscribing on walls or mon-
> uments was superseded by the use of papyrus and parchment. Books can be sold, sent,
> and given away...it was the portability and transmissibility of scrolls that brought the
> two nomadic tribes, first the Jews and later the Arabs, to replace the workshop of
> extremely heavy images of god with a god-given or even god-written book. The Bible
> and the Koran were only able to begin their victory march against all the temple stat-

ues and idols of the Near East and Europe because they were mobile relics. Because writing combines storage and transmission in a unique way, its monopoly held sway until media made letters and numbers and sounds technically mobile.[2]

Tracing the history of the book through its transformation from scroll to codex, miniaturization, and eventual mass production, it is apparent that as a communication technology books constitute the first and most enduring form of mobile media. With their flexibility, power and reach as a form of communications, books have become entrenched in elite, middlebrow, and popular cultures across literate sections of most societies. In a volume on contemporary mobile media, this chapter on reading provides an excellent opportunity to place cellular mobile phone technologies in historical, comparative perspective—and offer a critique of the myths and narratives concerning reading and this new media form.[3]

As mobiles become ubiquitous as a central form of contemporary media, reading itself has come sharply into focus. The cellular mobile phone ("cell phone") has been a remarkable personal and portable communications technology.[4] New forms of writing, textual communication, and reading were associated with cell phones from a relatively early stage, as we shall discuss. Since 2007, advanced multimedia phones (smartphones) allow different kinds of applications, programs, and affordances that push mobile media to the fore of e-reading. Popular apps on Apple's iPhone include readers, and with the popularity of tablet computers—notably the iPad, but also many other book-like, mobile computing devices—reading is a central function for users of mobile media devices. From another direction, the proliferation of all manner of e-readers—especially Amazon's Kindle, but Sony's long-standing E-reader, new devices such as Kobo, and the various e-readers being developed in China—is broadening the concept of mobile media. Such e-readers are utilizing mobile and wireless networks for the distribution and purchase of books, newspapers, and reading material and, as such, communication technology is being reworked by devices first designed as readers. The multi-functionality of the book is now more apparent that it has been for many centuries.

Against this backdrop, this chapter draws attention to the book as a cultural technology and connects this history to the emergence of the e-reader in the field of mobile media. It charts the history of e-readers, and when and how they begin to figure in mobile devices. The chapter discusses the early uses of mobiles for reading; a function that largely developed independently of e-reader technology plans. Then it moves to discuss the centrality of e-readers as an integral part of mobile media in the smartphone era. This is something that causes us to broaden our ideas of what mobile communication and media is, and complicates the politics, design, and direction of this eminently pervasive technology.

Book as Cultural Technology

The literary critic I.A. Richards once famously declared, "the book is a machine to think with."[5] His intention was to remind readers of the material and mental mechanics of the process of reading. Despite discussion of media "revolution" and "the end of the book," the reading formats and habits that have developed with the rise of mobile media do not represent a break with tradition. Books, on any platform, are still machines for thinking. Because of the apparent stability of the printed format of the book, it is easy to overlook it as a *communication technology;* a means for storing and retrieving information, subject to the same changes as any other media. Yet, since the fifth-century B.C.E. both the idea of "the book" and its practical function have been in a state of ongoing development.

The word "book" conveys a deceptive uniformity. For modern readers a book is something printed, or if ancient, written, on both sides of a sheet (parchment or paper) stitched or glued between protective covers. But, in its earliest form the book was a work inscribed on a long scroll. To be read it had to be deliberately unfurled and carefully held. The form of the scroll encouraged intensive reading habits—the close examination of one or two texts, often read aloud, and in groups, from beginning to end. These requirements made book use a specialized practice reserved for information of the highest cultural significance (largely religious texts). In contrast, the codex book with which today's readers are more familiar was designed with functionality in mind. Popular with lawyers, administrators and students, the codex could be used to record ephemeral data of only passing value. Wax-coated wooden tablets were linked together by string hinges, serving as "pages." The backs and fronts of each tablet were coated with wax that could be inscribed with a stylus and erased on the application of heat. Later versions of the codex technology featured washable parchment. Far from stable, these codex "books" were dynamic even by today's Internet-enabled standards. As opposed to their scrolling counterparts, these books were practical devices: with their string-hinged bindings they enabled users to sort and retrieve information in idiosyncratic ways. Their convenient size meant that individuals could use them independently.[6] Whereas readers of scrolls were expected to read the same thing again and again, the codex encouraged the habit of rapid and private reading. The defining feature of the codex was that it allowed for easier organization of information. With front-and-back pages, it could store twice as much data. The advantages of these features were unmistakable and the new religion of Christianity was the first to celebrate the codex's unique characteristics, elevating the format from a worker's tool to a communications network. Examining the early Christian codex bibles, Peter Stallybrass describes them as "the indexical computers that Christianity adopted as its privileged technologies."[7]

These historical developments in reading media demonstrate the fact often overlooked in present discussions about reading via electronic media: books have not always worked the same way with the same readers over the course of their history. As a media format the book is resistant to technological standardization. It adapts to suit the requirements of its readers and writers. As Robert Darnton points out, the needs and interests of readers in one era do not match those of another: "A seventeenth century London burgher inhabited a different mental universe from that of a twentieth century American professor."[8] Different cultures have used books differently over time, coming up with idiosyncratic functions in terms of use (reading left to right, up and down) and storage (pages or spines out). There is no reason to suppose that the latest innovations with the format and function of the e-book on smartphones and tablets represent any kind of "revolution," but rather, a continuation of this process of technical development.

Closer examination of the field of publishing demonstrates that the real "revolution" taking place is occurring at the level of industry where commercial concerns regarding the production and distribution of digital books are being explored, not always successfully. Too often discussions about the future of digital reading focus exclusively on the nature of virtual texts available for download "anytime, anywhere" subject to the whims of multi-tasking and impatient readers, but this overemphasis on the virtuality of texts adapted or created for networked devices obscures the many material realities to which acts of electronic reading and writing are subject. Richard Grusin rightly draws attention to the fact that digital reading is "dependent upon extremely material hardware, software, communication networks, institutional and corporate structures."[9]

Material constraints are often overlooked in favor of panic regarding the end of print and the triumph of digital networks, yet the technical history of the industrialization of the book offers a useful guide for understanding how e-books are equally subject to practical, material and commercial restrictions.[10] It was only with the industrialization of the printing process that the book's appearance and function came to be fixed in the manner we recognize. Until then the design, use and storage of books were highly variable. In the Renaissance, for instance, books served as domestic aides—used as filing cabinets and writing desks; taken as encyclopedias for notes, lists, clippings and diary entries.[11] Print technology streamlined the utilitarian function for the book and regulated how individuals read and understood the value of their texts. The misconception of the book as a fixed, permanent artifact of artistic rather than technological significance is undoubtedly connected to this industrial regulation. The process of industrializing publishing led to a popular reimagining of the book as a permanent monument to an eternal collective intellectual life. For example, the French writer Michel Butor complained in 1961 of all that has been lost since books were commercialized. Butor goes all the way back to Gutenberg to lament what has happened to books since they were ruined (in his view, almost literally) by technological advancement:

When the book was a single copy, whose production required a considerable number of work hours, the book naturally seemed to be a "monument" (*exegi monumentum aere perennius*), something even more durable than a structure of bronze. What did it matter if a first reading was long and difficult; it was understood that one owned a book for life. But the moment that quantities of identical copies were put on the market, there was a tendency to act as if reading a book "consumed" it, consequently obliging the purchaser to buy another for the next "meal" or spare moment, the next train ride.[12]

In Butor's opinion the older book is not a technology but a monument, an apparently "natural" form with obvious cultural worth in contrast to the "unnaturally" produced commercial paperback, for example. The newer book of multiple copies is insufficiently reverent to its content. Ironically, Butor can only make this claim about the book as a fixed monument by lamenting how the medium is subject to technological change. This tendency to normalize and de-technologize books has the effect of obfuscating the reality that, despite their ubiquity in our daily lives, books will always remain, in comparison to street signs or shopping lists, the most marginal of our reading and writing acts.[13] As Paul Erickson has noted, to regard the book as the "natural" form for information contradicts the central tenets of scholarship on the history of the book: practices of reading and writing go well beyond the book and, as a technology, books are not an unchanging medium for communication.[14]

When normalized or monumentalized, the book's place as a media format is ignored. This attitude also overlooks the opportunity to understand current practices of reading on electronic devices as part of a continuing practice of book production and consumption, rather than activities oppositional to print culture. A common but mistaken belief is that new technologies will always supersede or superimpose themselves on older media. This attitude ignores how technologies are actually lived and used. Early scrolls and codices existed side-by-side for hundreds of years, each serving the unique requirements and interests of their readers and creators. At various times the technology of the book has constrained and liberated readers; in response both readers and writers have taken various liberties, using the media to suit their own ends. By the 18th century, for instance, the discontinuous reading habits fostered by the codex were no longer new. Widespread literacy had opened up new markets for books. In response the new genre of the novel developed. In contrast to the usual discontinuous reading habit fostered by the codex, novelists created linear narratives that were dense with meaning, reference and symbolism. Contrary to all logic, except perhaps the logic of the market, the novel used a format wholly unsuited to its reading requirements, and came to make it its own. Despite serving today as a cultural touchstone for the value of the printed book, the novel is actually "a brilliantly perverse interlude in the long history of discontinuous reading."[15]

Scholars of new media now see similar experimentation and contradiction with reading and writing via mobile technologies. As we shall see, experiments with short-form episodic narrative composed via SMS are not only popular for reading on phones but have made it to the Japanese bestseller lists for print. Publishers are producing books-as-apps, seeing an opportunity to attract as many paying readers across as many platforms as possible. Novels on phones do not replace conventional print versions but rather complement them and offer readers an increased opportunity to engage with the text. All technologies involve constraint and liberation in equal measure. Thus, rather than thinking in terms of media revolution, positioning new and emerging reading habits in the context of the book's long and varied history of production and reception encourages a more nuanced understanding of the changing process of how we communicate—something necessary for understanding the emergence of electronic reading on mobile media.

Early Uses of Mobiles for Reading

While socially and culturally significant in many other ways, the arrival of portable voice telephony with mobile phones did not immediately lead to innovations in reading. Mobiles were a principally oral medium, and indeed their orality is still an important feature of their many hundreds of millions of users with low literacy levels. When this changes with the arrival of text messaging, as Leopoldina Fortunati evocatively suggests, it marks an "unexpected transformation, from the king of orality to a means of writing and reading."[16]

Something of an afterthought included in the second generation digital mobile GSM standard,[17] the first use of text messaging occurred in 1992.[18] From the end of the 1990s, text messaging became increasingly popular—especially among youth cultures.[19] SMS triggered a wave of user, design, and provider innovation centering on text communication, language, reading, and writing.[20] Among literary and artistic communities, there were many experiments in utilizing the capacities of short messages—for short fiction, poetry, and serialized work, and in new media art. Writers also used the capabilities of the mobile phone to create new kinds of epistolary fictions and interactive narratives. SMS was also widely used by other kinds of writers—such as those in the gaming community. Thus mobiles and text messages were incorporated into a wide range of games including forerunners of transmedia forms, in which reading works across a number of other inseparable modes of media ("platforms").[21]

The cultural and linguistic efflorescence supported by the affordances of SMS and its successor MMS seeded the most talked about endogenous development in mobile-supported literary fiction—the cell phone novel. Working within the 160-character limitations of SMS, epistolary, episodic, serialized, and other fictions were created by a continuum of amateur to professional writers. Japan was a pioneer, and in 2012 still can claim the most organized and salient publishing industry centering on cell phone novels. It appears that experiments with cell phone fiction commenced in Japan as early as 2000. Thus the emergence of cell phone novels was bound up as much with the pioneering Japanese mobile Internet[22]—and the famous i-Mode mobile data services ecosystem launched in 1999[23]—as it was with classic SMS. Cell phone novels were offered by the Japanese carrier responsible for i-Mode, NTT DoCoMo—constituting a niche market initially, compared to other mobile services such as ringtones and games.[24] The author "Yoshi" (a pseudonym) is credited with the first hit cell phone novel, *Deep Love: Ayu's Story* (2002), chronicling the life of a teenage prostitute in Tokyo.[25]

The existence of mobile community portals allowed users to post stories for download. In early 2007, the Maho i-Land ("Magic iLand") company established a free novel library.[26] The phenomenon became widely discussed in Japan, and the next year gained notice in the West when

> the [2007] year-end best-seller tally showed that cellphone novels, republished in book form, have not only infiltrated the mainstream but have come to dominate it....What is more, the top three spots were occupied by first-time cellphone novelists, touching off debates in the news media and blogosphere.[27]

In one of the few academic papers treating the subject, Larissa Hjorth argues that though cell phone novels were first the work of professionals

> ...by the mid 2000s everyday users had begun to be inspired to write and disseminate their own *keitai sh setsu*. Predominantly written by women for women, this mode of new media highlights the significance of remediation—many of the successful *keitai sh setsu* (millions produced yearly) are adapted into older media such as film, manga and anime.

The role of the Maho i-Land company was pivotal in this rise of the cell phone novel. Commenced in 1999, it provided an innovative matrix for various kinds of user productions, including do-it-yourself novels. According to Hjorth it was the "template 'Let's Make Novels'"—along with unlimited data packages for the *keitai* [Japanese word for cell phone] in 2003—that saw the dramatic rise of writers and readers of *keitai shōsetsu*."[28] Hjorth notes that "one of the key features of *keitai shōsetsu* is that they tend to follow a diary-like, confessional autobiographical model," and she suggests that the "intimate, personal nature of mobile media" rehearses "earlier epistolary traditions of women's fiction."[29]

While the Japanese cell phone novel experience has been widely noticed, there were other countries where cell phone novels attracted a significant following, including Korea, South Korea, and India. However, the other leading country most often attracting attention elsewhere for its SMS-based cell phone novels is China, where the form started circulating in 2004. SMS in China was very much associated with the new potential it offered as an alternative to official communication and publishing channels,[30] with new genres of popular expression such as SMS rhymes.[31] Cell phone novels become popular commercially[32]—indeed began to migrate to other media forms, generating new commercial opportunities. *Outside the Fortress Besieged* was a 4,200-character, 60-chapter novel delivered to subscribers in China via SMS.[33] In September 2004, the Taiwan-based Bestis Technologies bought the film rights to the novel, planning to produce it for viewing on mobile phones.[34]

So we can see evidence of extremely interesting and significant innovations in writing, reading, and literary production arising from the early decades of mobile phone. Such innovations operate at the intersection of industrial design and manufacture of phones, software applications development, innovation in everyday cultures of use, and changes in writing and literary production. They are distinctive to mobile phones and mobile media, clearly evident in the phenomenon of text messaging. Users themselves have played an important role in these developments, something that may be the most significant shift in practices of reading, as emerges from the case of the Japanese *keitai shōsetsu*. The other notable change in reading which is evident even in early experiments has to do with the implication of mobiles in other genres and form. Mobiles are widely used in multi- or transmedia, multi-site, collective experiences, evident in the implication of mobiles in games, art, locative media, and even digital storytelling, This is a potentially rich development in reading—and, down the track, the history of the book. We will take up this thread in the story of mobile media readers shortly, but now we will turn to an important parallel development that crosses over with mobile technology, but initially commences with ideas about computers, books, and reading.

Mobile E-Reading

The origin of the "dream of electronic books"[35] has been traced to Vannevar Bush's famous paper "As We May Think," where he proposes a device dubbed a "memex" that operates, like the human mind, by association.[36] One trajectory from Vannevar

Bush obviously leads through hypertext narrative and writing, and the World Wide Web—and now through the contemporary visions of social media. Another trajectory neglects the emphasis Bush puts on imagining a new associative technology, and rather realizes, even if only as an intermediate goal, the idea of storing and retrieving books. This later trajectory is largely what commercially available e-books have followed, since the term became widely used in the late 1990s. As Terje Hillseund explains:

> In the broad sense e-books have been around for several decades. In the Gutenberg Project thousands of books, mostly classic and public domain literature, have been made available for free as digital documents since the 1970s…[In 2001] the term e-book refers to digital objects specially made to be read with reading applications operating on either a handheld device or a personal computer. This modern concept of e-books came into common use after Martin Eberhart and Jim Sachs both started their own companies and developed Rocket eBook and SoftBook, the first two handheld e-book reading devices.[37]

Some commentators talk of the "first e-book reader generation," including the Sony Data Discman (c. 1990), the Franklin Bookman, as well as the early Rocket eBook and Softbook. The second generation—with "modem capabilities, greater memory, better screen resolution, and a more robust selection of available titles"[38]—included the SoftBook Reader 1998, Libruis Millenium device, and Everybook Dedicated Reader. In 1998, an organized effort began to create a common standard, with the Open eBook Structure produced the following year, precursor to the standards now produced by the International Digital Publishing Forum. With Microsoft involved in the open e-book standard, Bill Gates declared that:

> e-books promise to revolutionise the way the world reads. Whereas paper books are stand-alone entities, e-books can include hypertext links to additional content, whether it is in other books, databases or web sites.…In a paper book, content is fixed; with e-book technology it is flexible.…you will be able to get sound and moving images to support the text, creating an entirely new multi-medium.[39]

Progress was slow, however, and, and standards, particularly, remained a problem, especially with e-books, whereas some proprietary e-reader software, such as Adobe PDF, were gaining acceptable and portability across devices tantamount to de facto standards.[40] What was available was not inspiring either. In reviewing available e-books, no less than Alan Kay, pioneer of the Dynabook prototype of the early 1970s, judged that "little of what is good about books and good about computers was in evidence."[41] Things began to change when Sony launched the "Sony Librie," its e-book reader, the first to use e-Ink technology for its screen.

As the cell phone developed, it crossed paths with the evolving e-reader technology, and its associated industries, actors, and promoters. Cellular mobile phones were launched commercially in the late 1970s, and for their first decade and a half were principally associated with portable voice telephony and communications. With the availability of Windows Office, Adobe PDF, and other software on mobiles, especially smartphones—to compete with the portable digital assistant (PDA) market—more computer-screen-like reading on mobiles had been occurring. A French company, Mobipocket, founded in 2000, became the leader in e-book reading on mobile devices. Mobipocket is one of the three main e-book formats based on the Open eBook standards, focusing on offering book titles for reading on PDAs or a range of mobile devices operating systems (Windows Mobile, BlackBerry, Symbian, and so on). Acquired by Amazon.com in 2005, Mobipocket promoted its e-books under rubrics such as, "Did you ever try to read a book one handed?," "Did you ever try to read in the dark," and "Read everywhere."[42] While e-book software was available for mobile devices, and attracted a relatively small but enthusiastic following, it was more visibly promoted with Amazon's acquisition of Mobipocket. Clearly this deal meant that the fledging e-book software now had a bigger and wealthier backer. This piece of vertical integration also connected a bookseller with the application, and made the software much more attractive and easy to integrate for mobile phone companies.

The Finnish company, Nokia, a noted pioneer in seeking to cover all bases across mobile media and its potential applications, featured Amazon's Mobipocket as a key feature in the 2006 release of its flagship N95 mobile phone:

> The Nokia N95 comes out of the box ready to create, connect, consume and interact with some of the internet's most popular services. Use Yahoo! Search to search for and find most anything on the web, scroll though a book with Amazon's MobiPocket Reader or snap a photo and send it directly to your Flickr site.[43]

Under the tagline "It's what computers have become," the accompanying television advertisements extolled the virtues of the N95 as the ultimate cross-media device, with a veritable rainbow of culturally diverse actors intoning that "it would change my life, I wouldn't need to have three or four different devices, it's my magic box, my life, my future...it's my window—I can look through it, it's the world in your hand."[44] Notably, reading was not one of the featured activities in this ad. In one sense, this is odd, as during this period when mobile phones were being reconfigured as mobile media, there was scarcely an area where Nokia was not trying to establish itself as the convergent technology provider of choice—with dedicated, specific models of phones for each relevant domain of media culture: popular photography (high resolution camera phone), gaming (console-like device), television (larger screen), mobile computing (its early and acclaimed Nokia 9000i Communicator dating from 1997), mobile music, and mobile news and journalism.[45]

Certainly Nokia, like many other companies, researchers, and others developing and marketing mobile phone technology were very involved in designing screens for reading from at least the mid-1990s onwards—with much effort put into understanding how to best make text legible and manipulatable, especially through usability research into things such as scrolling, paging, browsing, layout, and text–image relationships.[46] Yet reading of books—as opposed to reading of text messages, application commands, maps, or mobile Internet browsing—is not an area of mobile media upon which Nokia, and indeed other mobile phone companies, appeared to focus. There were address books, phone books, travel booking applications, and browser bookmarks in early mobile phones—but not so many, or so explicitly, books as cultural technologies themselves. Of course, this changed dramatically with the advent of Apple's iPhone and the new age of the smartphone and tablet computer.

Tablet Politics: The Centrality of Reading in the Smartphone Era

Apple launched its iPhone device in mid-2007 to rapturous user response. Initially, however, reading and books again did not figure in its promotion. News and newspapers did feature early on, with advertisements depict browsing on the *New York Times* web site. Such an emphasis had quite a history, as news, particularly, had been developed for mobiles since mid-to-late 1990s—with alerts, information, messages with breaking headlines, premium mobile services, and, once better mobile Internet was available, reading newspapers via their online web sites.[47] Although Apple seemed to devote little or no effort to promoting e-books and e-reading, apart from news, no less than two years on, however, the iPhone was forging ahead in the e-book market, outcompeting Kindle and Sony e-Reader. One year later, the iPhone had emerged as a genuine force in the e-reader market.[48] It is unclear what Apple's corporate strategy for the iPhone, and various ideas about e-reading, were in the initial launch. Little or nothing is evident in the publicly available documentation. Rather what appears to have occurred is due to the iPhone, and especially its apps, as a platform, or arena, for innovation.[49]

Once Apple allowed third-party software development through the release of its Software Developers Kit (SDK), and launch of its Apps Store, various e-reader software become available. The two that initially become popular on the iPhone were Stanza and a piece of software with a pedigree on mobile devices, eReader (owned by Fictionwise). These pioneering e-reader iPhone apps allowed readers to tap into the existing stock of e-books in different formats that had been building up over some time. In February 2009, Google made available its public domain books through a mobile website for both iPhone and Android users—with readers using their browsers, as customary with Google Books. Bestselling titles were gradually made available via e-reader apps for the iPhone, including popular genres such as romance, fantasy, science fiction, and so on. In the meantime, "classics" featured heavily—as illustrated in the Classics app (with its stylized wooden bookshelf design).

No sooner had iPhone established itself as a handy medium of choice for many readers through the unexpected user-driver success of e-reader apps than the real battle in the economy of reading began. In November 2007, Jeff Bezos launched Amazon's much-awaited Kindle e-reader device. *Newsweek* called the event the "reinvention of the book," and "the first 'always-on' book."[50] Quickly adopted by readers in the U.S., the Kindle eclipsed the early-to-market Sony e-Reader, and became perhaps the first dedicated e-reader to gain wide adoption in the consumer market. This presented both a challenge and an opportunity for Apple's iPhone. On the one hand, Apple, by dint of the e-reader apps available for the iPhone, had become a very popular device for reading, echoing its entrenched position in games and music markets. Such was this popularity that Amazon's Kindle was forced to create an app for the iPhone to ensure that Amazon, rather than Kindle per se, was represented on the iPhone platform. On the other hand, the Kindle was a worthy, if not formidable, competitor to the iPhone—its larger format, customized for reading, was a preferred reading experience for many. And the Kindle opened the door directly to the vast trove of Amazon's book wares, quite an advantage given that Amazon was the leviathan of online book retailing. At the end of 2009, the bricks-and-mortar U.S. book giant Barnes & Noble launched its own e-reader, the Nook. Across the border, the Canadian Kobo e-reader launched in May 2010, a venture in which Indigo Music & Books holds the majority interest, and has partnered with Borders to offer the device in Hong Kong and Australia.[51] At the same time Kobo launched its software, which works on other e-reader devices, such as laptops, smartphones, and tablets—and allows access to a Kobo account for purchasing e-books, audiobooks, and other materials.

Mobile reading gained largely unexpected but impressive definition and momentum neither from the long-lived efforts of the e-book and e-reading developer communities noted above, nor from early proponents such as Sony or Microsoft—let alone the tepid forays of Nokia and other mobile technology companies. Rather it came alive from the new affordances, design, cultural appeal, and industrial might of computer and software giant Apple breaching the relatively closed field of mobile vendors and carriers with the iPhone. Mobile reading was also galvanized as a cultural phenomenon from the side of publishing and bookselling with Amazon's Kindle. As we have observed, by 2010, there were a multitude of devices, including a fleet of electronic and mobile readers in China that have been barely noticed in the West. Yet the most significant development still came from the reinvigoration of the tablet computer—in the form of Apple's iPad, launched in early 2010.

Just as the iPhone catalyzed competition and innovation in the smartphone market, so the iPad saw significant development, rebadging, and investment around the tablet computer—an easy-to-use and carry device, larger than a typical cellular mobile phone, but smaller than a laptop computer. Again like the iPhone, the iPad was not really a radical new invention; indeed it bore a much-remarked resemblance to an earlier form of "computer media"[52]—namely Alan C. Kay's fabled Dynabook from the late 1960s.[53] Rather, what marked the iPad out was a clever marshalling of many technical inventions and innovation in league with what might be thought of as "the Apple advantage": a globally recognized brand with a reputation for design; consumer acceptance in home computing and entertainment equipment, especially its computers and iPod; and a familiar content management and distribution system, in the form of iTunes.

From launch, Apple promoted reading as an integral part of the iPad. When Apple's founder, Steve Jobs, unveiled the iPad, he spoke glowingly of the iBooks app, claiming it as a breakthrough in e-reading. Apple deliberately targeted the digital publisher and bookseller community, and actively began to negotiate deals.[54] Unlike the iPhone, however, in which e-reading grew "unofficially," the development of the "official" iBook app and iBookstore was slow to develop, launching to coincide with the iPad. Because of its size, resolution, and compatibility with other Apple computers and e-reader software, the iPad has become the e-reader and document reader of choice of many of its early adopters but it is far from a versatile device: its operating system is not easily accessible, unless hacked (and so warranty voided); software can only be installed via the apps store; it has no ports; and available software for document reading is still clunky—though things have improved as apps have been refined, and with the launch of the iPad 2. Notwithstanding, iPads, Google Android-powered devices such as Samsung Galaxy Tabs, and other devices spanning the spectrum from tablet computer to smartphones are leading players in the transformations surrounding the cultural economy of reading. Mobile device-based readers are now the fulcrum of much-debated changes to the nature of books and reading, to the future of publishing and bookselling, and to the economic relations between writers, those who distribute books, and readers. Like online music before—a boon for consumers, but not necessarily for producers of music—e-books are now deeply involved in the recalculation of revenues and royalties across the existing publishing industries—with new kinds of business models, distribution channels, and artifacts themselves becoming possible.

Conclusion

More than ever before, books are machines for thinking and mobile media are presently being celebrated, fetishized, and reviled for their capacity to support and re-imagine the book. Above all else, and while we lack reliable figures on research regarding this, mobile media devices are being used for all sorts of mundane, as much as exotic, forms of everyday reading practices.

In this essay, we have argued that this striking contemporary emergence of mobile e-reading needs not only to be seen in the wider context of mobile media themselves; but also, critically, in the long histories of technologies, especially those pertaining to books and portable media. Thus we see a slow, complex emergence of reading on cellular mobile phones from the late 1970s until 2007, where the most obvious innovation arises from text messaging, and the intertwining of reading with other media and communication practices being remediated and extended with mobile media (especially narrative and artistic practices arising with mobile gaming and locative media).

By contrast, the serious investment in the development of specific new reading technologies such as e-books, e-readers, and electronic reading practices on mobile phones comes much later and is confined to a small minority. Mobile reading begins to flourish, rather unexpectedly, with the iPhone and then with the iPad, a tablet device that fittingly harks back to the confluence of mobility and media first enabled by scrolls, walls, monuments, ancient stone tablets and eventually the wax-coated codex.

Looking into the future, such histories provide us with resources for critique and reflection upon the significance of reading in mobile media, and its place in wider social and cultural transformations in which such technologies are profoundly implicated. While some leading mobile technology companies such as Nokia have been slower than others (for instance, Samsung) to embrace the tablet, and indeed to seize upon reading as a key future revenue stream for mobile media, we can point to signs that this is about to change. There are few reading apps or books that feature in Nokia's Ovi (apps) store, c. 2012. However, in early 2011, Nokia announced that Windows would henceforth be its smartphone platform.[55] While Windows Phone apps—through its Zune store—default to Kindle as its recommended e-reader, it can certainly build upon a long-standing investment in the area. This is obvious in its promotional material for its Windows Reader ("Read Anytime, Anywhere"), albeit still promoted as computer software, with an accompanying image of a woman reading using a laptop:

Why is Microsoft interested in electronic reading and eBooks?

Microsoft's vision is to empower people through great software—any time, any place, and on any device....Our goal with Microsoft Reader is not to replace the printed book. Rather, we hope to expand upon what is both useful and magical about books by offering a compelling new way to read. We also want to enhance the reading experience through features that take advantage of the unique benefits of the digital medium...[56]

Like many promises and myths of new technology, the reality of Microsoft's vision—along with the revolution of e-readers, Mobipocket's ubiquitous mobile reading, or Apple's desired breakthrough with iPads—has been underwhelming so far. Yet in other ways—in the experience of cell phone novels, or the interactive narratives created with mobile media, or the connected fictions and electronic literature emerging on mobile platforms, and earlier experiments in hypertext and networked arts—we can see great potential for mobile reading technologies to complement the many everyday acts of reading occurring in diverse places on a bewildering range of portable, mobile media around the world.

Notes

1. David M. Levy, *Scrolling Forward: Making Sense of Documents in the Digital Age* (New York: Arcade, 2001).
2. Friedrich A. Kittler, *Optical Media: The Berlin Lectures 1999*, trans. A. Enns (Cambridge, UK, and Malden, MA: Polity, 2010), 47–48.
3. See James Carey, ed., *Media, Myths, and Narratives: Television and the Press* (Newbury Park, CA: Sage, 1988); James W. Carey, with John Quirk, "The Mythos of the Electronic Revolution," in his *Communication as Culture: Essays on Media and Society* (New York and London: Routledge, 1989), 113–172; and Jeremy Packer and Craig Robertson, eds., *Thinking with James Carey: Essays on Communications, Transportation, History* (New York: Peter Lang, 2006).
4. Outlined in the classic study of Japanese *keitai* (mobile phone), namely Mizuko Ito, Daisuke Okabe, and Misa Matsuda, eds., *Personal, Portable, Pedestrian: Mobile Phones in Japanese Life* (Cambridge, MA: MIT Press, 2005).
5. Ivor Armstrong (I.A.) Richards, *Principles of Literary Criticism* (London: Routledge & Kegan Paul, 1926).
6. The practicalities of the codex took on a symbolic dimension when the early Christians decided to use the form as the chosen medium for their version of the Bible. Jewish communities wrote their sacred scripture on scrolls and by moving to the new codex form—with its associations with daily life—it is thought the Christians may have been visibly

demonstrating their break from their Jewish roots. This "break with tradition" in the ancient world resonates with current changes to communication technology that have seen new developments in portable, personal media. See Peter Stallybrass, "Books and Scrolls: Navigating the Bible," in *Books and Readers in Early Modern England,* ed. Jennifer Andersen and Elizabeth Sauer (Philadelphia: University of Pennsylvania Press, 2002), 42–79.

7. Stallybrass, "Books and Scrolls," 74.

8. Robert Darnton, "What Is the History of Books?," *Daedalus* 111, no. 3 (1982): 78.

9. Richard Grusin, "What Is an Electronic Author? Theory and the Technological Fallacy," in *Virtual Realities and Their Discontents,* ed. Robert Markley (Baltimore: Johns Hopkins University Press, 1996), 46.

10. Among many corrective are contributions to Hans Ulrich Gumbrecht and K. Ludwig Pfeiffer, eds., *Materialities of Communication,* trans. William Whobrey (Stanford, CA: Stanford University Press, 1994); Joseph Tabbi and Michael Wutz, eds., *Reading Matters: Narrative in the New Media Ecology* (Ithaca, NY: Cornell University Press, 1997); and Adrian Mackenzie, *Cutting Code: Software and Sociality* (New York: Peter Lang, 2006).

11. Jeffrey Todd Knight. "'Furnished' for Action: Renaissance Books as Furniture," *Book History* 12 (2009): 37–73.

12. Michel Butor, *Inventory: Essays,* ed. and trans. Richard Howard (New York: Simon & Schuster, 1968), 42.

13. D.F. McKenzie's essay on bibliography draws our attention to how marginal the book really is in the history of our textual communication and how mutable has been its physical form. See D.F. McKenzie, *Bibliography and the Sociology of Texts* (London: The British Library, 1986). See also Ann R. Hawkins, ed., *Teaching Bibliography, Textual Criticism, and Book History* (London: Pickering & Chatto, 2006).

14. Paul Erickson, "Help or Hindrance? The History of the Books and Electronic Media," in *Rethinking Media Change: The Aesthetics of Transition,* ed. David Thorburn and Henry Jenkins (Cambridge, MA: MIT Press, 2003), 95–116.

15. Stallybrass, "Books and Scrolls," 47.

16. Leopoldina Fortunati, "The Mobile Phone: Toward New Categories and Social Relations," *Information, Communication & Society* 5(2002): 513.

17. Friedhelm Hillebrand, ed., *GSM and UMTS: The Creation of Global Mobile Communication* (Chichester, UK: John Wiley, 2001). On SMS and MMS, see Gwenaël Le Bodic, *Mobile Messaging Technologies and Services: SMS, EMS and MMS* (Chichester, UK: John Wiley, 2005).

18. Friedhelm Hillebrand, ed., *Short Message Service (SMS): The Creation of Personal Global Text Messaging* (Chichester, UK: John Wiley, 2010), 126.

19. Eija-Liisa Kasesniemi, *Mobile Messages: Young People and a New Communication Culture* (Tampere, Finland: Tampere University Press, 2003).

20. See Richard Harper, Leysia Palen and Alex Taylor, eds., *The Inside Text: Social, Cultural and Design Perspectives on SMS* (Dordrecht, The Netherlands: Springer, 2005); and David Crystal, *Txtng: The Gr8 Db8* (Oxford: Oxford University Press, 2008).

21. Christy Dena, "Transmedia Practice: Theorising the Practice of Expressing a Fictional World across Distinct Media and Environments" (PhD thesis, University of Sydney, 2010).

22. On mobiles and the Internet in Japan, see the classic volume *Personal, Portable, Pedestrian: Mobile Phones in Japanese Life,* edited by Mizuko Ito, Daisuke Okabe, and Misa Matsuda (Cambridge, MA: MIT Press, 2005).

23. On i-Mode and Japan's pioneering mobile services eco-system, see accounts by two of its creators: Mari Matsunaga, *The Birth of i-Mode* (Singapore: Chuang Yi Publishing, 2001);

and Takeshi Natsuno, *The i-Mode Wireless Ecosystem,* trans. R. S. McCreery (Chichester, UK: John Wiley, 2003).

24. Associated Press, "Cell Phones Put to Novel Use," *Wired,* March 18, 2005, http://www.wired.com/gadgets/miscellaneous/news/2005/03/66950.

25. Patrick W. Galbraith, "Screen Dreams: A Digital-Age Literary Form Has Become a Metropolis Powerhouse," *Metropolis Magazine,* January 22, 2009, http://metropolis.co.jp/features/feature/screen-dreams-2/.

26. Lisa Katayama, "Big Books Hit Japan's Tiny Phones," March 1, 2007, http://www.wired.com/culture/lifestyle/news/2007/01/72329.

27. Norimitsu Onishi, "Thumbs Race as Japan's Best Sellers Go Cellular," *New York Times,* January 20, 2008, http://www.nytimes.com/2008/01/20/world/asia/20japan.html.

28. Larissa Hjorth, "The Novelty of Being Mobile: A Case Study of Mobile Novels and the Politics of the Personal," in *Throughout: Art and Culture Emerging with Ubiquitous Computing,* ed. Ulrik Ekman (Cambridge, MA: MIT Press, 2012).

29. Hjorth, "The Novelty of Being Mobile."

30. Zhou He, "SMS in China: A Major Carrier of the Nonofficial Discourse Universe," *The Information Society* 24 (2008): 182–190.

31. Haiqing Yu, *Media and Cultural Transformation in China* (Abingdon, UK: Routledge, 2009), 75ff.

32. S. Sangwan et al., "Mobile Communications Market in China," in *China Information Technology Handbook,* ed. Patricia Ordóñez de Pablos and Miltiadis D. Lytras (New York: Springer, 2009), 306.

33. Dan Steinbock, *The Mobile Revolution: The Making of Mobile Services Worldwide* (London: Kogan Page, 2005), 89.

34. Steinbock, *The Mobile Revolution,* 277.

35. Larry Press, "From P-books to E-books," *Communications of the ACM* 43, no. 5 (2000): 17–21.

36. Vannevar Bush, "As We May Think," *Atlantic Monthly* July 1945, http://www.theatlantic.com/magazine/archive/1969/12/as-we-may-think/3881/4/ (accessed February 15, 2011).

37. Terje Hillesund, "Will E-books Change the World?," *First Monday* 6, no. 10 (2001), http://131.193.153.231/www/issues/issue6_10/hillesund/

38. Nancy K. Herther, "The ebook Reader Is *Not* the Future of Books," *Searcher* 16, no. 8 (2008): 26–37.

39. Bill Gates, "Beyond Gutenberg," (1999), http://www.microsoft.com/presspass/ofnote/11–19billg.mspx.

40. Philip Barker, "The Future of Books in an Electronic Era," *Electronic Library* 16, no. 3 (1998): 191–198.

41. Alan C. Kay, "A Review Article: Dynabooks: Past, Present, and Future," *Library Quarterly* 70, no. 3 (2000): 385–395.

42. "Why eBooks?," Mobipocket, http://www.mobipocket.com/en/Corporate/About Mobipocket.asp?Language=EN (accessed February 15, 2011).

43. Nokia, "It's What Computers Have Become—The New Nokia N95," media release, September 26, 2006.

44. Nokia, "Nokia 95: It's What Computers Have Become," advertisement, http://www.youtube.com/watch?v=vmSmdDJ4SQg&feature=related.

45. On Nokia's approach to mobile phone culture and mobile media, see Gerard Goggin, *Cell Phone Culture: Mobile Technology in Everyday Life* (Abingdon, UK: Routledge, 2006) and *Global Mobile Media* (Abingdon, UK: Routledge, 2011).

46. For instance, see Christian Lindholm, Turkka Keinonen, and Harri Kiljander, eds., *Mobile Usability: How Nokia Changed the Face of the Mobile Phone* (New York: McGraw-Hill, 2003).

47. For a discussion, see Gerard Goggin, "The Intimate Turn of News: Mobile News," in *News Online: Transformation and Continuity,* ed. Graham Meikle & Guy Redden (London: Palgrave Macmillan, 2010).

48. "Apple's iPhone Could Become Next Hot EReader Says Report," *eWeek,* November 2, 2009, http://www.eweek.com/c/a/Mobile-and-Wireless/Apples-iPhone-Could-Become-Next-Hot-EReader-Says-Report-633757/.

49. See Harmeet Sawhney and Seungwhan Lee, "Arenas of Innovation: Understanding New Configurational Potentialities of Communication Technologies," *Media, Culture & Society* 27, no. 3 (2005): 391–414; and Gerard Goggin, "Ubiquitous Apps: Politics of Openness in Global Mobile Cultures," *Digital Creativity, 22*(3), September 2011, pp. 148-159.

50. Steven Levy, "The Future of Reading," *Newsweek,* November 17, 2007, http://www.newsweek.com/2007/11/17/the-future-of-reading.html.

51. Susan Krashinsky, "Kobo Takes a page from the Kindle," *Globe and Mail,* March 25, 2009, http://www.theglobeandmail.com/report-on-business/kobo-takes-a-page-from-the-kindle/article1511433/.

52. On computer media, see Paul A. Mayer, "From Logic Machines to the Dynabook: An Overview of the Conceptual Development of Computer Media," in *Computer Media and Communication: A Reader,* ed. Paul A. Mayer (New York: Oxford University Press, 1999), 3–22.

53. For the conception of the Dynabook, see Alan C. Kay, "A Personal Computer for Children of All Ages," in *Proceedings of the ACM National Conference* (Boston: Association for Computing Machinery, 1972), http://www.mprove.de/diplom/gui/Kay72a.pdf.

54. Calvin Reid, with Jim Milliot, "Apple iPad Invades Digital Book World," *Publishers Weekly,* February 1, 2010, http://www.booklife.com/pw/print/20100201/41878-apple-s-ipad-invades-digital-book-world-.html.

55. Microsoft, "Nokia and Microsoft Announce Plans for a Broad Strategic Partnership to Build a New Global Mobile Ecosystem," media release, http://www.microsoft.com/presspass/press/2011/feb11/02–11partnership.mspx.

56. "General Questions," Microsoft Reader, http://www.microsoft.com/reader/info/support/faq/general.aspx.

8

As It Happens

Mobile Communications Technology, Journalists and Breaking News

COLLETTE SNOWDEN

Another day, another disaster. Or so it seems. From floods, earthquakes, tsunami, tornadoes and volcanic eruptions to elections, riots and revolutions on the streets, news and information have never been transmitted as quickly from where it's happening to where you are. Much contemporary news is comprised of a constant stream of information about emerging events, often overlapping in their intensity. Just as rapidly, however, they "fall off" the attention radar of audiences and media as new events emerge. The news cycle has always been like this, but the speed at which a news event rotates through the news media and the consciousness of the audience has accelerated.

Breaking news about events appears in traditional news media and on online and mobile news sites in much the same way as it has for decades; that is, as news breaks the media begin to report it as soon as they can, providing they consider it important enough. But there is a significant difference now. Much of the content received as breaking news, "as it happens" or "this, just in," is no longer produced by journalists. Instead, breaking news material frequently appears as an event occurs, close to or occasionally in real time, and as an event unfolds the amount of material available proliferates. The range of means which allow people to receive news gives more people immediate access to eyewitness accounts, stories, images and audio reports of events as they occur, often from multiple vantage points and perspectives. This "reporting" is the work of ordinary people located where something interesting, amazing, or important happens. In the past victims or spectators to an event might have accessed the media by being interviewed by a journalist covering the event; now the use of mobile communications technology provides them with immediate access to a global networked distribution system. Victims and spectators are able to tell their own stories, and to record their own reactions, feelings and perceptions, often before mainstream media organizations can mobilize a journalist or camera crew to generate a report. Material is not only available, accessible and distributed, but begins streaming online and into the public domain long before the first journalist appears to cover events. User-generated content, material produced by users of communication technology, is challenging the traditional role of journalists in providing public access to news as soon as it happens. This development is largely a result of the diffusion of mobile communications technology.

The move to user-generated content via mobile communications technology has been faster than anticipated, even by optimistic forecasters, and with each new development or refinement of the technology such content increases. The multi-modal connectedness to the global communications network that mobile communications technology provides with apparent ease is at the heart of this development. The narrowly controlled media environment that has been under pressure as the communication technology has developed has seen the privileged role of the media usurped, and in some cases rendered irrelevant. Whereas journalists were usually the first to report events, or to bring news of momentous events to a mass audience, they now compete for attention with a diverse range of "reporters" who record events using compact mobile digital devices.

The use of mobile communications technology is now so widespread that it has disrupted the conventions governing the public distribution of news and information. These conventions had developed over decades, and in many respects were fundamental to definitions and understanding of journalism, both as a profession and as a practice. The reporting of news as quickly as possible has been an essential and prized characteristic of journalism and the news media since it evolved.

> The first recorded commercial news service was initiated in the 1830s by Daniel Craig and his carrier pigeons. Traditionally, early U.S. newspapers sent agents out in boats to intercept ships from Europe; they collected news items and quickly returned to specified points ashore, from which riders on horseback carried the news to the respective papers. Demonstrating considerable ingenuity and enterprise, Craig took homing pigeons with him to the incoming ships; when released, the news-laden pigeons outraced the horsemen every time. This news was sold to all interested parties....When financial considerations were absent, this speed became more a matter of pride among journalists.[1]

The ability to be first with the news, including getting a "scoop" or exclusive story, i.e., material that no other news outlet or journalist has been able to obtain, was and continues to be highly valued by journalists and an attribute that contributes to their professional identity. Örnebrin argues that "existing journalistic work practices have grown out of a need to adapt labor to this use of speed as the main measure of competitive success in the news industry."[2] However, the professional identity of journalists and the work of journalism has been progressively challenged and undermined by the increasing access of non-journalists to communications technology and the means to distribute news and information without the need to access established media.

Shoemaker and Reese's theorization of a hierarchy of influences on media content argues that among the key factors affecting the production of content by journalists is the self-definition of the work of a journalist. They argue that

> it seems clear that the way in which journalists define their jobs will affect the content they produce. Journalists who see themselves as disseminators/neutrals should write very different accounts of an event than those who see themselves as interpreters/participants.[3]

The work of journalism has been radically redefined by the blurring of responsibility for the production of content between journalists and audiences, particularly in breaking news situations, where events are unplanned, unexpected, in hard to access locations, or on a large scale. The introduction of content produced by "non-journalists" has displaced and subverted the conventions and practices that have developed for the reporting of breaking news. These conventions were largely based around the special privilege afforded to journalists and their capacity to gain access to an event of significance or interest. Historically, this was not possible for anyone other than those individuals who had resources available, including transport and access to the technology necessary to report via the distribution system of a mass media organization. Journalists were granted special privileges and status to do this work. In the past, even the most erudite and courageous witnesses could not share accounts of events or experience unless they gained access to the media distribution system, and these accounts were generally rendered in the third person. The low cost and high convenience of mobile communications technology has placed an effective still and video camera, text generator, audio recording device and the means to broadcast in the hands of hundreds of millions of people globally. At the same time telecommunications networks provide immediate access to a global distribution system, which negates the need for mediation by established media organizations.

People's willingness to use the technology is understandable, especially when "always on" functionality and ease of use are considered. Hermida argues that these qualities contribute to the presence of "ambient journalism," that is, journalism, which was once difficult and expensive to produce, today surrounds us like the air we breathe. Much of it is, literally, ambient, and being produced by professionals and citizens.[4]

Frosh and Pinchevski define the outcome of this process as "media witnessing" and argue that it "refers simultaneously to the appearance of witnesses in media reports, the possibility of media themselves bearing witness, and the positioning of media audiences as witnesses to depicted events."[5] They argue that the role of mobile communications technology is significant for the phenomenon of media witnessing. Subsequently, other scholars have adopted the concept to analyze the more nuanced and detailed consequences of the process of witnessing events as they happen through the media and through a mass shared, mediated reporting of an event, accessible globally. Frosh and Pinchevski also argue that "these new constellations of recorded experiences defy traditional models of mass communication, producing ad hoc communities of attention on a global scale."[6]

The witnessing of events afforded by the use of mobile communications technology has at its base the firsthand testimony of ordinary people, who, Turner argues, are becoming increasingly visible in the media.[7] Significantly, a deep understanding and familiarity with the received and accepted conventions of the media and the news media enables the construction and delivery of that testimony via the use of mobile communications technology. Generations of people now have experienced media content as consumers and audiences and understand its format, formula and requirements at least well enough to mimic them. The transition to user-generated content was therefore relatively seamless, because users understood the conventions of media reporting. That is, as consumers and audiences of media, they knew what the content should look and sound like. "Performing" as a journalist was not difficult for many people, and it was assisted by the willingness of audiences to accept the technical and qualitative imperfections of user-generated content, often where it came from mobile sources, in a trade-off for authenticity and immediacy. This has been especially so where the content clearly originates from mobile sources.

Yet, the producers of this content are not necessarily "citizen journalists,"[8] in the sense that in many cases there is no conscious decision or intention to report events. While material produced by "citizen journalists" for websites and blogs with the intention of providing alternative or independent perspectives is deliberately produced, much of the breaking news that mobile technology enables and provides is serendipitous and uncalculated. In many cases the material is produced simply because people turn on the camera or video function in their mobile phone, or call or text someone, or make a post to a social network as they watch and/or experience an event as it happens. For many people, turning on a mobile device in unusual circumstances appears to be a reflexive and expected action. Whatever means are used, and for whatever reason, the production of this material is circumventing and disrupting the almost exclusive role that journalists have had historically as the reporters of breaking news. The traditional role of journalists as the recorders and reporters of events, either as on the spot witnesses to events, or as reporters dispatched to provide news in the immediate aftermath, is challenged directly by the production of material from accidental or incidental witnesses using nothing more than their digital mobile devices.

The proliferation of user-generated content, individually produced, widely distributed, and freely available, has also disrupted the capacity of government and corporations to control or censor the reporting of news or information about events. When images, stories and updates of events are distributed without filtering, and in real time, the capacity to control the flow of information is also limited. The crudest attempt to censor the flow of information, closing down telecommunications systems, is more damaging. As Arianna Huffington argued in 2009 in response to the action of the Chinese government after riots in the western province of Xinjiang,

> China just delivered a stunning, real-world demonstration of the changes rocking—and transforming—modern journalism.
>
> When deadly riots broke out in the western province of Xinjiang earlier this month, the Chinese government sprang into message control mode. It choked off the Internet and mobile phone service, blocked Twitter and Fanfou (its Chinese equivalent), deleted updates and videos from social networking sites, and scrubbed search engines of links to coverage of the unrest. At the same time, it invited foreign journalists to take a tour of the area.
>
> That's right, it slammed the door in the face of new media—and offered traditional reporters a front row seat.
>
> China's leaders realized that it's one thing to try to spin the on-the-ground views of bused-in reporters ("To help foreign media to do more objective, fair and friendly reports," in the words of the government's PR agency), but quite another to try to spin the accounts and uploaded images of tens of thousands of Twittering and cell-phone camera-wielding citizens.[9]

The inability for news and information to be contained and controlled as easily as in the past is indicative that the democratization of content production promised in the early days of the internet has eventuated.

Modern mobility (accessible and fast transport) and communications technology are the twin pillars that have facilitated the transmission of information and allowed the delivery of breaking news to be a task undertaken en masse. The journalistic task of "bearing witness," which in previous generations required the resources and financial support of media organizations, has thus become the work of participants and spectators. In the event of wars, political riots, social revolutions,

terrorist attacks, and in natural disasters such as tsunami, forest fires, floods and tornadoes, it is common practice for people to use mobile technology to capture what
is happening around them, to describe it and then distribute it widely. Such is the
scale of the practice, and our acceptance of it, that we no longer have to wait for news
media sources to publish or broadcast material. User-generated content flows unfiltered and raw into the "mediasphere"[10] through social networking sites, through
which it spreads virally. Frequently, the task of the traditional news media and journalists is to obtain and share the most interesting or best quality of this material and
appropriate it for their own use, before their competitors do so.

The ways that news is received and produced in the second decade of the 21st
century therefore differs considerably to the news reporting of the early 20th century. Warnings about the potential for change as communications technology has
developed a range of new platforms and its consequences for journalism have been
discussed consistently alongside the technology. Yet, the scale and speed of the
change associated with online and mobile communications technology has been
consistently underestimated by journalists and media organizations. As a profession
journalism also has not successfully responded to the change, despite its history of
early adoption of technology and its capacity to adapt professional practices to
accommodate new forms of technology.

In 2002 the future of "peer-to-peer news" was discussed by established commentators, including Justin Hall, who speculated, "Perhaps we will see the same
flourishing of amateur reporting and video experimentation witnessed after the
release of the Sony PortaPak in the mid-sixties," and "Perhaps some businesses will
host wireless weblogs."[11] One event changed such speculation into reality—the
Boxing Day Tsunami in the last week of 2004. When the tsunami struck it was during a period of the year around the world when newsrooms were sparsely staffed.
The lack of available "on-the-ground" journalists coincided with a major holiday
period, a massive event spread across multiple locations in 55 countries, and major
disruption to transport. Immediate access to the many different locations affected
by the tsunami was difficult and restricted. However, before journalists could even
find their passports, on-the-spot reports, and telephone accounts were available from
hundreds of witnesses and survivors as they reconnected with their friends and families using telecommunications. Almost immediately the media began looking for
the content produced. Images and video footage became available rapidly, much of
it derived from tourists who watched the tsunami and filmed it as it struck their
beach resort hotels. Some of the material, however, including footage of water and
debris surging through streets in Banda Aceh, was not seen for two weeks or more.

One of the legacies of the Boxing Day Tsunami was a realization that the media
could not accurately report the scale of the disaster. Even the largest media orga-

nizations found it difficult to mobilize staff and technical resources. The traditional journalistic role of reporting from the scene of the disaster was undertaken within existing journalistic conventions; that is, there was a focus on reporting details of what happened, what was being done, stories of survival and promoting the need for assistance. However, this traditional reporting was heavily supplemented by the flow of stories, images and videos that began to appear from spectators and survivors.

Only a few weeks later the arrival of the video-sharing website YouTube irrevocably changed the practice of reporting breaking news by making it possible for people to upload and distribute a wide range of media content, without requiring it to be "mediated" via traditional media organisations.[12] By the time of the London Underground bombing in July 2005 the capacity for the media to obtain useful, often powerful material, produced on mobile phones by spectators and participants of events, became even more obvious. A week after the bombing *The Guardian* reported:

> It was a new kind of story. Not in the sense of what happened, which was thoroughly and depressingly as anticipated, but in the way it was reported and disseminated. The mobile phone photographers, the text messagers and the bloggers—a new advance guard of amateur reporters had the London bomb story in the can before the news crews got anywhere near the scene.
>
> Emerging from inside the police cordon, ordinary tube travellers brought out dramatic footage that defined the media coverage, leading the evening TV news bulletins and staring out from the pages of the next day's newspapers.[13]

The 2005 "London Bombing" has been intensively studied and analyzed by media scholars, and it is widely considered to constitute a "tipping point" in the transformation of both mobile media content production and the response to it by traditional news media organizations. From this time the use of participatory media content was no longer considered speculative or marginal to the work of the "real" media. Instead, it was clear that mobile user-generated content had effectively evolved to become a powerful and influential media practice. Writing a month after the bombing, Wallace summarized the impact of the deluge of mobile user generated content,

> Even if it was an overstatement to describe the reporting of the July events in London as a revolution in communication, as many mainstream media commentators did, it certainly was a bold announcement that the reporting of major events would never again be the exclusive preserve of the mainstream media. The amateur reporter had made an entrance and would be a permanent guest at the table where the first draft of history is written.[14]

From then on the use of mobile-communications-distributed, user-generated content has permeated the media, not just in the United Kingdom, but globally.

Journalism

The mass generation of media content has affected journalism in a number of critical areas, notably, journalistic procedures and practice, professional identity, and in relation to audience interaction and participation. However, the focus here is on the way that mobile communications technology has changed, and is changing, the processes and practices of journalism, from which other aspects of professional transformation necessarily flow, as Hardt[15] and Pavlik[16] have argued.

One of the most critical attributes of mobile communications technology is the immediacy with which it enables content to be distributed. In comparison, even the best forms of traditional journalistic practices are inadequate, except perhaps for live radio and television. However, live broadcasts require a presence that must be either planned or possible. In many cases mobile communications technology is able to provide reporting, on the spot or as it happens, or at least provide a record of an event before the media become aware of it. The inability of journalists and media organizations to compete with this attribute of the technology leads to such practices as public pleading for material. This is evident in the practice of media participation in social media sites, not as producers of material, but as consumers. While the news media have historically solicited material from the public and encouraged the audience to act as informants about situations and events, this practice has escalated.

For example during an earthquake in Christchurch, New Zealand in 2011 (the second major earthquake in the city in six months), a number of messages from media organizations appeared on the social media micro-blogging site Twitter. For example, Australia's SBS radio and television network news staff tweeted,

In Christchurch? Please let us know if you have any information or pictures.[17]

While the Australian Broadcasting Commission tweeted a similar request:

ABC TV News looking for pictures or people to talk to us about the latest #chch#nz#quake, please get in touch.[18]

The Australian TEN television network also appealed for information and images:

Large quake hits Christchurch. We're starting a @Storify with latest reports http://bit.ly/lbT9x7 Tweet us if you have info, images.[19]

The reference to Storify points to a further development in the way that the news media are responding to multiple sources of information in breaking news situations. Storify is an example of how the amount of user-generated content is being managed and how journalistic processes and practices are being modified in response to the content available. It is a platform that aggregates material from multiple sources, especially social networks, which can then be reordered, or explained, or contextualized. Storify is promoted as "a future content management system for social media."[20] Its potential to provide a focus for the streams of information now being produced meant that it was quickly adopted in its Beta form by various traditional media organizations, such as the *Washington Post*, but also by online media sites, such as the Huffington Post.[21]

A significant feature of Storify is that it was developed by a former Associated Press reporter, Burt Herman, who argues that

> Everyone can be a "reporter" when an event happens. But not everyone is a "journalist"—making sense of an issue and giving the context. So we built a system to help people do this, take the best of social media and make it into a story—to "storify" it. The word itself is actually in the dictionary, and also comes from my AP days when editors would send messages to bureaus asking them to "storify" something.[22]

The focus on the aggregation of content and its contextualization, as emphasized by Storify, is not exactly a new practice, but rather the intensification of an existing one, combined with the use of new technological platforms. Whenever large or significant stories have broken, it has been the practice of news media to allocate staff and resources to "cover" them, a process which involved and required the location and aggregation of information. With the task of "covering" breaking news becoming larger, and requiring a constant, real-time response, the development of a technological solution is a rational response. It is particularly so given that the trend toward a greater emphasis on breaking news has occurred at the same time as many news media organizations have reduced the numbers of journalists they employ, in response to declining audiences and income.

The greater emphasis on the role of the news media as an aggregator, rather than an originator, of information is also occurring more frequently. This is apparent in the establishment of practices by the news media to focus on updating information as it emerges and is fed to the media, often by audience members using mobile devices. This has produced the practice of "rolling" news bulletins or live blogs, often updated minute by minute, which track the progress of events, and provide links to related articles, images, opinion pieces and commentary.

Mobile communications technology facilitates the provision of news and information from multiple locations and perspectives, but also calls attention to the need for centralized organization or "sense making" of the material produced. Again, however, this is not exactly a new practice, but rather the extension of an existing practice. From the earliest days of the telegraph, news organizations were able to receive reports of breaking news. Many news rooms had their own telegraph receivers and operators, and were linked to the newswire companies, which had changed and shaped journalism. This was one of the critical transformations of journalism through the technologization of its processes and practices.[23] Indeed, the influence of technology on reporting and breaking news, and the relationship between technology and being first with the news, became more commercially significant with each new development in communication technology.

While Craig's carrier pigeons focused on delivering news quickly over space, the medium of radio allowed news to be broadcast in real time. Radio's capacity to communicate breaking news, regardless of distance, had an emotional power that reached its zenith when understood and exploited in the infamous *War of the Worlds* broadcast by Orson Welles, in which the reportage "passed from one eyewitness voice to another as the story is told in the form of a newscast gone awry."[24] Heyer also notes that Welles drew heavily on the famous reporting of the crash of the airship Hindenburg by Herbert Morrison, which although recorded live was not broadcast until the following day.

Other forms of technology have also been used by the media to provide immediacy and give audiences access to news as close to real time as possible. For example, police radio scanners have been a staple technology in newsrooms for decades, allowing journalists to "follow" the activities of mobile police units and respond to events deemed significant. This use of technology saved journalists and their employers both time and resources. Television news uses running news banners or "tickers" to provide breaking news updates—without interrupting scheduled programming; originally the ticker was used to report current stock prices, effectively taking the traditional stock ticker from the newsroom directly into the media. The "ticker" was adapted by 24-hour news channels to broadcast breaking news and news headlines, and quickly became a common feature of television news broadcasting. Existing media organizations are therefore following the pattern of the news media throughout its existence in adapting and developing processes and practices to incorporate new technology, including those associated with mobile communications. However, whereas previous generations of communications technology have enhanced the role of journalists and contributed to the perception of the profession of journalism as privileged and special, mobile communications technology is not following that pattern. While mobile technology allows journalists to report and contribute to breaking news stories, as well as to use individual micro-blogging and social media accounts via mobile, they do so in a much more crowded communication environment. In that environment, many non-journalists have equal access, and sometimes better access, to the audiences that have traditionally been the target and the market for journalism and its special commodity, news. In effect, the widespread availability and use of mobile communications technology, with all of its media production and distribution capability, has challenged the monopoly that journalists have had on the production and distribution of news.

The capacity to adapt quickly and creatively to developments in technology has therefore become a requirement for professional survival in journalism. Multi-skilling is now demanded, rather than desired. However, Deuze argues that studies of journalists internationally demonstrate that as a profession journalism is fractured and confused by the requirements of rapid adaptation to change and the disruption occurring not only to the practices and processes of journalism but to the understanding of what it means to be a journalist.[25]

An account that provides insight into the practice of contemporary journalism is found in the blog of Canadian journalist, Erik Rolfen, digital news editor at *The Province* newspaper, writing after riots broke out in Vancouver following the Canucks' loss to the Boston Bruins, in the National Hockey League final for the Stanley Cup in 2011.

> There's nothing quite like arriving in a newsroom when the shit's starting to hit the fan. If you're unaware that news is breaking, it becomes quite clear before you've taken three steps inside. People are moving at twice their normal weekday pace, and shouting across cubicles has displaced all other forms of communication.
>
> The next five hours were a blur. On the web desk, we each quickly found a niche; Katie working the social media accounts, Cheryl compiling reporters' dispatches into coherent stories, Dharm attacking all the video footage that came in. I manned the live blog that we had set up at the top of our home page, pushing out our reports, photos and videos, and anything worthwhile that our eyewitnesses or competitors were coming up with, in as close to real time as possible.[26]

This event also produced an enormous amount of material. But it was a powerful image, captured by a professional news photographer, "The Vancouver Kiss," that garnered global attention. In the image

> we see a young couple laid out on the street, embraced in a kiss, juxtaposed between a blurry riot cop running towards the camera and a line of police charging a crowd in the background. Once the photograph hit the internet, it became a viral phenomenon in minutes. Already the uncontested frontrunner for the photo of the year, it's being compared to other historical kisses like Robert Doisneau's "The Kiss" and Alfred Eisenstadt's picture of a kissing couple in Times Square on VJ day.[27]

Debate about the authenticity of the image, and whether it had been staged, quickly followed its publication. The debate, however, was answered just as quickly by the emergence of additional material, both still images and videos captured from other perspectives on mobile devices, which showed that the couple had, as they said in interviews, been knocked down by riot police.[28]

The audience as a producer assists the media production process, but the audience also has the potential, through the use of mobile communications technology, to assume a more active role. There is not only a willingness to share, but a desire to. People are keen to show what is happening to them. There is also a compulsion to indulge in competitive display in the sharing of material, almost with the message "If you think that's awesome/terrible/shocking—have a look at this." This practice has become so common that the amount of material produced is frequently beyond the capacity to be absorbed or accessed by individuals. In 2011 when a tsunami devastated parts of Japan, the amount of material, mostly produced using mobile devices, quickly became overwhelming. One of the critical factors this event showed, seven years after the Boxing Day Tsunami, was how quickly the practice of recording events, even dangerous and life-threatening ones, has become normalized. However, as Winston argues, the unintended consequences of media[29] are often the most enduring, and here the consequences of saturation coverage of news by the media became apparent. Our capacity to witness, as Frosh and Pinchevski[30] have defined it, may be limited by what we can actually bear, and we reach a point where we turn away. Consequently, the increasing news coverage that mobile communications technology allows increases the speed at which we reach the point where the need to turn away surpasses the need to know what is happening. Consequently, our capacity to pay attention and focus on events in our environment is thereby reduced, rather than enhanced.

The recording of abnormal events or situations has developed as an almost intuitive, reflexive response so that people record events, not only because of the circumstances, but regardless of those circumstances. News media are no longer the sole or dominant provider of breaking news. As journalism has been required to do throughout its existence, it is adapting its processes and practices in response to the changes associated with technology. The media no longer has a commanding control of the flow of information about the present and has, in many areas of reporting, lost its role as gatekeeper and agenda setter. As a consequence journalism, and the news media generally, are reshaping priorities and re-ordering the practices and processes associated with the defining and delivery of news. Now that much of the content that was once "exclusive" to the media, and both copyrighted and commodified, has become globally accessible, there is both an economic and professional imperative for the media to control new ground in the mediascape. The relevance of the news media is dependent on the development of a new model, even as the environment constantly changes around it.

Consequently, the media frequently assumes the role of anticipating and projecting the future. Rather than reporting news "as it happens," the media begins a process of speculation and projection ahead of incidents and events, in which opinions proliferate and reports are produced in which experts are asked to explain what *might* happen, what *could* happen, what *will* happen. This practice may be regarded as "anticipatory news." Rather than objectively witnessing and reporting events, journalists are more frequently assuming the role of interpreter. Unable to be first with actual reporting, they instead attempt to be first with commentary or analysis. Consequently, there is a greater emphasis on warnings, reminders of historical events, contextualization and speculation by the media. In the service of the projection of the near future, the media is reconstituting itself and engaging in a form of role reversal where journalists provide a post-event interpretation, rather than engaged, eyewitness accounts.

However, the media is also better placed to report on the consequences of events, especially natural disasters. After the Japanese earthquake and tsunami in March 2011, in an interview with Bruce Shapiro from the DART Center for Journalism and Trauma, Irwin Redlener, M.D., founder and director of the National Center for Disaster Preparedness at Columbia University, argued that the media had to be unrelenting in its reporting AFTER major events, particularly natural disasters.

> The first thing to understand it that every time a major disaster occurs anywhere in the world we rush to label it a wake-up call. In fact, they are not really wake-up calls. We get aroused, we focus our attention, the media is covering it 24/7; we're awed, we're distressed, it's terrible—but then we move on. The news gets to be repetitive, the stories don't really have a broader meaning; and then it's off the front pages with an occasional popping back up if something new is going on.
>
> What we don't do is take the lessons out of these big disasters and then apply them to make us more prepared for, or more resilient to, future disasters. I call them "snooze alarms"—we just go back to sleep, back to the complacency.[31]

In this role the media is better placed to report on the consequences of events, especially natural disasters. When the urgency of the surveillance function that creates the initial surge of interest in an event has passed, the need for analysis and dissection of the causes and consequences increases rather than diminishes. But our focus on emerging news is allowing this role of journalism to be diminished and to be subsumed by the need for speed and a focus on emerging and breaking stories. Rather than writing the first draft of history, the media is better situated to become an editor and authenticator of the ongoing narrative. This is a substantial change of role, which requires equally substantial changes in how media practices and processes are perceived and managed, in the understanding of the identity and role of the journalist in society, in reporting conventions and in the relationship between journalism and its audiences.

Now that every form of human activity is accessible to those with access to the technology and the knowledge to navigate it, we are losing the capacity to be shocked or surprised by what is openly displayed. Instead we are shocked and surprised by secrecy, by the suppression of information, by the ability of people to deceive, to hide, or to corrupt, often in plain sight. If journalism means more than simply conveying information wrapped in entertainment, if the special role of journalism is as the Fourth Estate, as the profession itself has argued so loudly for so long, then it must uncover secrets, reveal deception, discover what is being hidden and reveal corruption. It cannot do that sitting in an office, protected by security guards, waiting for news to flow into an inbox, or be delivered by a stream of social media or micro-blogging updates. In order to do that journalism must become an active profession, rather than a passive, clerical job. It may even need to discard its emphasis on speed, and focus more on the slower and painstaking search for accuracy and truth.

Notes

1. Stephen Shmanske, "News as a Public Good: Co-operative Ownership, Price Commitments and the Success of the Associated Press," *The Business History Review* 60, no. 1 (Spring 1986).
2. Henrik Örnebrin, "Technology and Journalism-as-Labour: Historical perspectives," *Journalism* 11, no. 1 (2010): 57–74.
3. Pamela J. Shoemaker and Stephen D. Reese, *Mediating the Message: Theories of Influences on Mass Media Content*, 2nd ed. (White Plains, NY: Longman, 1996), 101–102.
4. Alfred Hermida, "From TV to Twitter: How Ambient News Became Ambient Journalism," *M/C Journal* 13, no. 2 (May 2010), http://journal.mediaculture.org.au/index.php/mcjournal/article/view/220 (accessed June 30, 2011).
5. Paul Frosh and Amit Pinchevski, eds., *Media Witnessing: Testimony in the Age of Mass Communication* (New York, NY: Palgrave Macmillan, 2009).
6. Paul Frosh and Amit Pinchevski, "Media Witnessing and the Perpetual Ripeness of Time" (lecture, February 1, 2011), http://histcon.se/file/89/media-witnessing-lecture-february-1–2011.pdf
7. Graeme Turner, *Ordinary People and the Media: The Demotic Turn* (London: Sage 2009).
8. Shayne Bowman and Chris Willis, *We Media: How Audiences Are Shaping the Future of News and Information* (Reston, VA: The Media Center at the American Press Institute, 2003).
9. Arianna Huffington, "Bearing Witness 2.0: You Can't Spin 10,000 Tweets and Camera Phone Uploads," *Huffington Post,* July 13, 2009, http://www.ihavenet.com/China-Bearing-Witness-You-Cannot-Spin-Tweets-and-Camera-Phone-Uploads.html
10. Arjun Appadurai, "Disjuncture and Difference in the Global Cultural Economy," in *Global Culture,* ed. M. Featherstone (London: Sage, 1990), 295–310.
11. Justin Hall, "Mobile Reporting: Peer-to-Peer News," *The Feature* (blog), February 20, 2002, http://thefeaturearchives.com/topic/Media/Mobile_Reporting__Peer-to-Peer_News.html
12. David Croteau, "The Growth of Self-Produced Media Content and the Challenge to Media Studies," *Critical Studies in Media Communication* 23, no. 4 (2006): 340.
13. The Guardian, "We Had 50 Images within an Hour," July 11, 2005, http://www.guardian.co.uk/technology/2005/jul/11/mondaymediasection.attackonlondon
14. Milverton Wallace, "A Citizen Journalism Identity Emerges from London Bombings," Newsstand, UCLA Center for Communications & Community, August 2005,

http://www.c3.ucla.edu/newsstand/global/a-citizen-journalism-identity-emerges-from-london-bombings/

15. Hanno Hardt, "The End of Journalism: Media and Newswork in the United States," *The Public* 13, no. 3 (1996): 21–41.

16. John Pavlik, "The Impact of Technology on Journalism," *Journalism Studies* 1, no. 2 (2000): 229–237.

17. @SBSNews (Twitter 13 June 2011, 13.11.52 P.M. ACST)

18. @ABCNews24 (Twitter 13 June 2011, 13.47.42 ACST)

19. @channelten (Twitter 13 June 2011, 13.14.45 p.m ACST)

20. Storify, http://www.storify.com

21. Kenna McHugh, "What Is Storify and Why Did They Raise $2m?," *SocialTimes* (blog), February 9, 2011, http://socialtimes.com/what-is-storify-and-why-did-they-raise-2m_b37661

22. Burt Herman, cited in Kenna McHugh (February 9, 2011), http://socialtimes.com/what-is-storify-and-why-did-they-raise-2m_b37661

23. James W. Carey, "Technology and Ideology: The Case of the Telegraph," *Prospects* 8 (1983): 303–325.

24. Paul Heyer, *The Medium and the Magician: Orson Welles, the Radio Years, 1934–1952* (Lanham, MD: Rowman & Littlefield, 2005).

25. Mark Deuze, "What Is Journalism?: Professional Identity and Ideology of Journalists Reconsidered," *Journalism* 6 (2005): 442–464.

26. Erik Rolfsen, "It Ends with a Whimper and a Whole Lot of Bangs," *Erik Rolfsen* (blog), June 18, 2011, http://erikrolfsen.tumblr.com/post/6649634701/it-ends-with-a-whimper-and-a-whole-lot-of-bangs

27. Douglas Haddow, "Vancouver's Kiss of Life," *The Guardian*, June 17, 2011, http://www.guardian.co.uk/commentisfree/cifamerica/2011/jun/17/vancouver-kiss-rioting

28. Raw Live Video of Vancouver Riot Kissing Couple, http://www.youtube.com/watch?v=TenWb-xLqDE

29. Brian Winston, *Media, Technology and Society. A History: From the Telegraph to the Internet* (London: Routledge, 1998).

30. Paul Frosh and Amit Pinchevski, 2011

31. Bruce Shapiro, "Japan Aftermath Demands 'Unrelenting' Reporting," Dart Center for Journalism & Trauma, March 19, 2011, http://dartcenter.org/content/uncharted-territory

9

Time and Space in Play

Saving and Pausing with the Nintendo DS

SAMUEL TOBIN

The Nintendo DS, a series of handheld game systems, has been ignored, with a few exceptions, by cultural critics and game researchers. Perhaps the DS, small in stature, has simply hidden in plain sight? For whatever reason, the DS has simply not attracted researchers' and scholars' attention compared to, for instance, the current generation of powerhouse home game consoles or PC facilitated Massive Multiplayer role-playing games. However, the DS deserves our attention. It is hugely popular and near ubiquitous with 132,040,000 units sold worldwide as of this year (Nintendo consolidated sales report), it is an elegant expression of mobile and miniature technology, and, as I will explore in this chapter, it can be a pathway into fundamental questions concerning play's relationship to time and space.

The DS allows its users to play with storage, transmission, and communication, key issues for any study of media, as described by Harold Innis in his epochal *Bias of Communication*.[1] Innis's whirlwind history of the shifting fortunes of ancient empires and the role of communication media in those fates is a classic of North American media studies. This work was particularly influential in its focus on the temporal and spatial aspects of technologies, and their tendencies to be oriented towards permanence and duration over time or speed and domination of space. I draw on Innis's work here to show how if we think through the uses of the Nintendo DS as being about these same fundamental modes of transmission and storage, of stability and fragility, we can understand video game play and portable video game play in new and important ways. We can do this, keeping Innis in mind, by paying close attention to how DS players save and pause their play. By attending to pausing and saving, we can arrive at a thicker, more adequate description of play, and by engaging with these non-diegetic activities that are generally considered to be external to game play, we can begin to address critical aspects of portable video game play practices that might otherwise be overlooked. This article's aim is to show how the DS can and should be studied in relation to the larger goals of media, sociology, and game studies, for in the use of this miniature and mobile device, we can see how play redefines space and time.

To that end, I draw on a heterodox group of texts from classic sociology to media studies as well as current work on video games in order to support my argument and to illustrate aspects thereof. DS play is described in detail in this article with the first section focusing on what can be shared and communicated through the public use of the DS. The second section draws on Erving Goffman, in particular his treatment of materiality and play in his dramaturgical sociology to draw out the potentiality of play to transform space. The third section traces the relationship between saving and playing to explore how the DS players engage with time through pausing and storing play. The article concludes with an exploration of what is at stake politically and ethically in the modes of play addressed throughout the piece.

Transmitting in Space-Based Media

The Nintendo DS, which stands for "Dual Screen," is a portable video game system, which has, in one iteration or another, been manufactured by Nintendo Co Ltd since 2004. There are (so far) five versions of the Nintendo DS. They are, in order of release, the DS, the DS Lite, DSi, DSi XL and the most recent 3DS. While many models make up the series, for the purpose of this chapter I will refer to them generically as a group, as "the DS." Although this approach runs the risk of glossing over details, it is common practice amongst the players as well as journalists and other video game commentators to use the term DS to refer to all models. More pragmatically, the differences between models (cameras, size, etc.) do not affect the behaviors and usages of players (saving, pausing, etc.) that are the focus of this chapter.

The DS was (and still is) trumpeted for many features, one of which is its ability to connect, wirelessly, to other units via Wi-Fi, allowing partnered or team play, competition, and the sharing of games as well as the "picto chat" function. This online functionality is increasing and approximately a quarter of owners of the

Nintendo DS have played at some point online.[2] No doubt this trend will continue. And whilst the majority of DS players still do not use the Wi-Fi capabilities of their DSs, they are, in effect, still engaged (even in "offline" mode) in transmission and storage, in playing with and through time and space. I approach this device in its offline mode in order to explore in detail aspects of its use that its Nintendo Wi-Fi networked play options otherwise may occlude. To that end, my focus here is on another kind of connectivity offered by the DS, specifically on the kind of interactions and the manner of *transmission* in casual play with the DS. Even without using the "wireless" feature, DS players interact and communicate with others, who, whether they like it or not, share in some way in the play. When we bring our DSs out and about, when we play in public space, we perform our play for people who have varying degrees of interest, comprehension, and attention to our activity. To better understand this, I will draw on Erving Goffman and some Goffman-inspired work done on the dramaturgical and framing issues found in the use of mobile phones in public.

To invoke Bernard Suits,[3] what we share, what we show when we play in public with our DSs, is a *lusory attitude;* we announce with our bearing that we are bent on play. While the DS's design and use may overlap with many other portable electronic devices, particularly Palm Pilot-type organizers that have a stylus, the DS is recognized as a device used primarily for games and for play. With new iterations of the iPhone being more game friendly and with Sony selling intellectual property or content through Universal Media Discs, this distinction between phones, game-playing devices, and organizers is rapidly losing meaning. But for now, it is still clear that the user of a DS is a player and the use of the DS in public is therefore a performance of play, a display of a lusory attitude. What do we transmit when we play with the DS? First and most importantly, we convey with our public DS play that we are in fact playing and that the space we share with our onlookers is a space that whatever its other functions is now, at least for the moment, also a site for play. To appreciate the significance of this play, we need to blur the boundaries between who is playing and who is present, who is watching, merely tolerating, or not noticing at all. As James Newman writes, "videogames are not exclusively solitary experiences, regardless of what popular discourses might suggest about their inherent asociality."[4] Newman argues for a more expansive role of what he calls "Off-Line" players or second players, by which he means people who without a controller in their hands nevertheless watch, help, and play, which is a different level of engagement than that of onlookers, fellow travelers, and people in line who share, willing or not, the play space created by the public use of the DS. I am extending Newman's concept to make a more elastic definition not just of whom a player is but what it means to engage with or have a relationship to video games. Just as Newman used the concept of "Off-Line ergodicity" to move beyond the controller-in-hand-pressing buttons as delimiting the definition of a game player, I want to move further to a kind of area-effect reading of public play, a reading in which there is no room for asociality or, in fact, where asociality is just a specific kind of sociality.

In her work on how space and affect are managed and transformed by public use of cell phones, Chantal de Gournay[5] suggests that cell phones allow users to draw in, announce, and otherwise signal something about their own emotional and social state and status. For de Gournay a mobile phone user's speech is directed not just at a listener on the other end of phone call but for people nearby who, willing or not, overhear that speaker. What is key here is the public-ness of the use of the cell phone and the way this public use creates an overlapping of spaces. Public cell phone use communicates something about the state of the user to people who share public space with them. While it might at first seem a stretch to compare the communicative power of mobile phones, which after all are defined as communication devices, with the communicative power of the DS, which is defined as a personal game-playing device, the comparison is useful. Signs warning against the use of music players and cell phones in travel spaces are an institutional acknowledgment of the potential of these devices to engage willing or unwilling others who share space with the caller, listener, or game player. There are many other forms of play, some non-technological, that can be performed in public, engaging, annoying, signaling, and announcing contested versions of the meaning of "public" in public spaces as well as what kind of space they are. Playing in public announces that any given space, whatever its other uses, is now, and potentially always, one of play. There are, of course, also other portable media, both old and new, which can be played with in public places, including "travel-sized" board game and playing cards, perhaps one of the most evocative of ludic mobile technologies.

The Prisoners' Ball

In this section I focus on the kinds of communication practices we participate in when we play in public, both communication with past and future versions of ourselves through saving, pausing, and loading games, and also the way we communicate and perform with these devices in the presence of others and proximity to others. Here we might think of contingent spaces such as waiting rooms, and any room in which we wait.

The question is what is at stake in these situations? What kinds of habits and systems are we required or asked to cultivate, engage with, tolerate, and even enjoy in order to occupy and move through these spaces? What does it mean to experience time and space in this particular manner, one which we may either strive to make playful, recast, or transform, or one which we suffer through and under, as governed by regimens revolving around work, which we go to, return home from, and/or try to avoid? Erving Goffman[6] suggests a more hopeful (and contradictory?) reading of play. Goffman's reading of games and how they make irrelevant or later some aspects of material life (for example, a coin's monetary value is irrelevant if it is to be flipped for "heads or tails" or if it is to be pitched against a wall in a game) is one that is important for my larger argument, one that is centered on a material practice, the social-technical power of tools or toys to change space and time.

If we push this further, we can approach what is at stake in DS play not merely from the prospective of exploring the explicit potential opportunities of social game play, as for example through the DS's wireless potential or from a reading of a DS game's diegetic level, and instead look at how irrelevance flattens some material functions of the DS while transforming others. Consider how the irrelevance of materiality in game playing works not just to smooth over differences of wealth between the players but also to redefine space and circumstances, as can be seen in Goffman's example of prisoners' ability to play "wall games," in which within the duration of the play the prison (or school, or mental hospital), wall becomes a game space rather than an imprisoning enclosure.[7] Play shifts focus and meaning, re-contextualizing and recasting time, space, and relationships.

Stylus Traces

Not only can play recast space through the evocation of alternative realities, play can rework time through pausing, saving, and recording evidence of play. Play on the DS toys with time through its very materiality. For instance, the DS's touch screen bears a record of past play on its surface. Like many owners of the DS, I purchased a thin plastic transparent screen or slip-cover to protect the bottom touch-screen from enthusiastic use. If we remove the prophylactic screen and hold it up to the light, we can see a flurry of scratches, indentations, and loops and whorls made by the stylus as well as the smudges and the oily residue of our fingers and thumbs. Unlike the palimpsest of Archimedes, we can uncover here no written language, but that does not mean there is nothing here to be recovered. How should we think about a palimpsest that lacks words?

What are these smudges and scratches? Evidence? Information? Records of exertion, of use? Yes, but of what kind of use? Are these specifically evidence of play? The sweat stains and the grime are traces associated not with ludic activities but with activities belonging to the broader category of things we use with our hands, things

that can become well-worn through this use and attention. This reading emphasizes the materiality of not just the thing, but also of the relation of the thing and
its user. The blister on the user's fingers testifies to just how hard it has been pushing against the surface and the interface of the thing; the sweat stains and scratches
on the surface in turn show with what effort the thing has been marked by the user.
The traces on the slip screen are like a tiny version of an ice rink after the skaters
depart. What the scratches on the screen protector show is a digital record, not in
the electronic sense of digital, but rather in the tactile sense, that someone was here
and they did something here with a tool, in this case not with a skate, but with a
stylus. This is a scribbled-over map of play-time.

The stylus, in concert with the touch screen, sets the DS apart from not only
other handheld game systems but from games systems in general and is at the same
time that which connects it with other handheld devices, such as some types of palm
pilots. That the "touch screen" can be used without the stylus is also important as
this is increasingly (post iPhone) an important model for HCI design. The stylus
and the finger are both addressed by Walter Benjamin[8] in a passage in which he discusses the typewriter and its relation to the hand and to the (receding) fountain pen.[9]
While Benjamin is talking about writing, typing, or otherwise creating a text and
not "merely" playing with these tools, the key here is his explicit description of the
change in input device as being a tactile, manual issue and that different ways of
interacting with mechanical systems can be both mimetic—performing/referenc-
ing the finger and hand—while also innervative in Benjamin's particular sense of
the term, which is to incorporate these gestures as technologies.

The stylus of the DS works in a number of ways. Much of the loading and game
selection screens that a player must go through when the unit is first turned on are
operated by the touch screen.[10] These activities can be done with the stylus, but in
my experience, particularly when used in short bursts of casual play, it is much more
likely that one would use a thumb than pull out the stylus for these tasks.[11]

The stylus also works in the save capacity in more than one way. First, pro-
saically but importantly, in some DS games there is a function by which a player's
avatar can open a notebook, in which she can, using the stylus, write notes about
the game, often a hint or another piece of information that must be remembered
for later use; for example, a location of a hidden item in *Legend of Zelda: The
Phantom Hourglass*[12] or a code in *Retro Game Challenge*.[13] Of course, a pen and paper
can work in much the same way (an interface that was "built in" via blank "note"
pages at the back of many old NES manuals); it is important to remember that the
use of a pen and paper is less likely when playing with a DS in transit.

At another, more abstract level, the DS stylus allows players to save. What is
it that the stylus stores or saves, if not the data on its attendant cartridges? The fact

that this data, the files of our game play, are saved to the game cartridge and not to the (operating) system suggests that we need to look for other ways in which we save with the DS. The stylus, though not the only interface option, works as both an index and symbol of this potential towards a more open-ended perspective and treatment of saving or storage could be/could mean in videogames and in game studies.[14] Is it useful to expand our conception of record(ing) to include signs of use? If these marks are readable, what can we read in them beyond proof of labor and struggle? The computer memory records in-game achievement, but not the sweat and muscle invested. It is a question of digital actions or residue versus digital recording. This is a matter of the accrual of bruises and scratches produced by physically effortful and even damaging palimpsestic play, leaving a patina which links the DS to that range of artifacts which achieve the status of "well-used." This evidence of material effort acts as record not of specific in-game events but rather of the vitality of the player and the tactile, corporeal relation a player has with their DS or for that matter that a user may have with any mobile device in this anthology.

Storage and Time-Based Play

The Nintendo DS, like practically all modern portable video game systems (excepting the classic/primitive Nintendo Game & Watch series and other dedicated one game systems), allows a player to save a game and then to continue it later, or else to try a tricky part of the game again. However, what is interesting about the DS use in this regard is how saving in the diegetic in-game mode, which writes directly to the cartridge, is not as important in general as the ability to *pause,* which is practically universal in portable video game play.

It is not merely by code and program that we find our play saved. Surely our progress through a game space or narrative or procedure is saved and stored at the level of data. However, we do save when we use the DS in other ways as well. Certainly we also save something in the form of experience, memory, and acquired skills, but my focus is on how we save these experiences materially, in and on the device itself, the object, not the code, as well as in our own bodies.

Effort, work, and bodily activity are all stored, but not in data, but rather in the material aspects of the DS and the DS assemblage. This kind of conceptualizing of storage forces us to reevaluate what it is that is stored or saved. I will finish my exploration by arguing that we need to account for the evidence of interactions and uses that can be read through traces, strains, and wear and tear across the assemblage from palimpsestic scratches to worn buttons and callused thumbs. By animating this kind of storage and saving and in attending to it, I hope to map out not only a more robust conception of what playing with these things is like, but also what is at stake, raising larger questions about media, storage, and transmission.

We should also consider that players at times stop playing without formally pausing, which would require closing the DS or pressing "start" in the middle of a game. This kind of action is risky in terms of preserving play because in most games there are only a few instances open to such moments of respite and relaxation. J. P. Wolf[15] discusses these events, conceptualizing them as part of the game but extradiegetic in that they are not moments when the player is actively engaged in gameplay. He calls these moments "interludes," and "moments in which the game's interactive potential is briefly suspended," which he attributes to the mechanics of the device, as for example moments of stillness that occur when the program's code is loading new information or refreshing a screen, or moments designed to provide a modulation of tension and a sense of "pace" to the game, or "breathers" for the player as well as scripted death animation sequences or avatars' victory dances.[16]

Load screens, pause buttons, item inventories, and other non-deigetic moments allow the player to sit back, stretch, and otherwise withdraw from the game-posture and from the demands of time and speed. The controller works differently in these non-diegetic contexts. In the innovatory or "settings" screen, the controller becomes mouse-like. The game play pauses (literally) for data-management; the interface owes much to a Windows or Mac desktop as well as to the old high-score screen of the arcade. The setting screens may not be part of the play, but they are still part of the game. The "load" screen works differently. With the "load," or between moments, the screen comes into play (or not) as the system brings up the next stage, "level," or area of the game. In these moments the controller as the interface between the player and the game is nervous, even superstitious. A and B buttons are worked like worry beads. This is the game play version of flicking your fingers under the faucet waiting for the cold water to get hot. The commands given here go nowhere (they go where the typing between active windows goes). Something happens but nothing registers; you are talking to yourself.

Harrison Gish[17] suggests that these moments, which usually come at the beginning or end of a level of play, are usefully charged with information, and therefore are part of playing the game well, if not actually what most would call "game play." My interest, however, is in what the player can do in these moments beyond this kind of game-space reconnaissance, and this is where Wolf's "breather" is particularly useful.[18] We use these moments in any kind of game play situation to attend to things, our person (can one ever need to scratch one's nose more than in the midst of heated game play?), and otherwise take care of something else, but briefly. If we think about situations of public portable play (as opposed to the more relaxed domestic arrangement of the couch and console), particularly in transit as in a subway, we also may need a second or two to let someone by or to shift our weight. While these kinds of pauses are outside of the game, they are all about our relation to the play we are trying, however precariously, to inhabit. These pauses merit our critical attention as they can help us to see that we pause in order to save play.

To pause is to save play from the specter of labor, from completionism, from instrumentality, and, taking a term from RPGs (a popular genre for the DS), to save us from the dubious pleasures of "grinding." The play that pausing saves is casual play. The save function in contrast has more to do with the kind of play that starts to feel more and more like work. Connecting these practices back to Innis's ideas about time- and space-based or biased media, saving is concerned with permanence, pausing with flexibility. Saving (Innis's time-bias) is a matter of building and accrual, of options, paths, tricks and best practices as well as of progression through a game narrative, points or levels. As Wolf puts it: "With the save function, players could…save a game at a point where a decision was required, and then play the game out multiple times from the saved game, making a different choice each time."[19] Play oriented towards saving ought to be worth saving and play in this sense tends to be careful. Pausing (Innis's space) is orientated towards the casual and its play is fast and loose. The pause act stretches play, give it breathing room,[20] a hesitation rather than a restart. Pausing is for when we don't know for how long we can play. Pausing is associated with movement even if it is only anticipated movement, while saving-oriented play is associated with stability and having the time and space for a good game session worth storing.

Pausing has no promised history like saving does, no potential outside of play itself and play's ability to fill up the wait of the waiting room, to make more bearable the coach seat, the lecture, the train platform. Given the contingent nature of the kind of play I examine here, it is a matter of minutes not hours, or if hours, then hours made of one more minute at a time. In this casual play, nothing is promised, and what is stored is not meant to be archived. While the high score exists, and it may be reached in this sort play, that would be incidental, at best a happy surprise but not the inspiration to open the DS. The difference is one of duration, intent, and lusory attitude. Playing for high scores, beating the game, and wanting to complete it are ways of playing for keeps, having more in common with the kind of playing that we plan for, work toward, and, to quote David Sudnow, about which we must "care." We save this play to build off of later.[21.] This kind of play is certainly important with the DS: the popularity and success of the RPG genre is testament to that. It is important, however, that we not only account for what the device affords vis-à-vis what is possible in terms of technical specifics of memory, storage, and power, but also how the DS affords and makes possible an alternative kind of management of time, and along with this, a different kind of play style or ludic attitude, one that is based not on the capabilities of the device to store and retrieve data permanently but of a player, either by shutting the DS's case, hitting pause button or even just waiting for a lull in diegetic action to hesitate, wait or pause. What the DS and its pause feature allow is a kind of redemption of play as category of action, that is, in Goffman's sense of term, irrelevant.[22]

Play Time

I have endeavored to keep my focus on the almost toy-like character of the DS and to offer a material and embodied as opposed to a game-oriented reading of play on and with the DS. Thus far, my focus has been on the DS's life in our hands: how we touch it, open and close it, and carry it around with us. In combination with this attention to the material aspects of DS play, I need to address the workings of time, an aspect of play with the DS that goes beyond my discussion of pausing.

In Goffman's example, prisoners' playing with their bouncing balls transforms the prison wall into a goal, net, or backboard, and in this way we see how play works to redefine, recover, and recuperate space. However, the other key dynamic in Goffman's ballgame is that the play situation is temporary, lasting only as long as the prisoners are playing; as soon as play stops, the backboard reverts to being a prison wall. The other time-based dilemma that the prisoners face is the time they spend in confinement—the duration of their play occurs against the backdrop of the much larger time frame of their prison sentence. At the risk of being overly dramatic, I would suggest that this is the situation in which we play with our DSs, a situation not of a prison but of empty time spent in depot waiting rooms, airport boarding areas, subway platforms, and in transit. The ethical and political problem we face when we describe play this way as transformative, even liberating, but only temporarily, is that no real freedom can be contingent, and yet this is where we are left. If we need to kill time, it is because that time is arduous and oppressive. Maybe there is the possibility for a more lasting, freer, transformative play. But the situation I describe here, in which the DS functions as a stopgap, a time filler and time killer, as a palliative technology, is not to say that DS play is not legitimate, germane, or important. The prisoners would admit that the ball was not going to free them, but that doesn't mean that they didn't want or need it.

With this article, my intention has been to contribute to the evolving critical discussion of play as a concept, category, activity, and ideal, a discussion that is still as yet under-theorized. People are playing a lot, and a lot of what they are playing with are handheld devices. Over 330 million games were sold in the past year for the Nintendo DS, and while raw masses of numbers are not reason enough to study something, they do point toward a gap between what gets written about in game studies and what is actually played.[23] By focusing on a new reading of play attuned to such key concepts as time, experience, and space, we can open up fundamental questions about pleasure, bodies, technology, interface, and the subject.

I have tried to account for and unpack how users of the Nintendo DS save and pause their play by examining such practices through frameworks of time- and space-based media and the materiality of that play. I have reworked Innis's concepts of storage and transmission in a number of ways. First, by showing how the DS allows us to transmit not just through explicit networked connectivity but also in the sharing of lusory and ludic attitudes and social transformation of space. Secondly, by arguing that we save and store with the DS not just at the level of data and files but also on the surfaces of the object itself. Finally, I have shown that using the pause function of the DS is often in response to the exigencies of the space the player is in rather than some event in the game they are playing.

By examining pausing and saving in the context of the DS, we expand and sharpen our understanding of what constitutes play with these handheld game systems. I have argued that we need to define video game play as including these extra-diegetic yet still ludic practices of saving and pausing, of dealing with time- and space-based constraints and affordances. These play practices, despite their minute scale (mere moments really), are filled with tensions, decisions, and interventions into fundamental issues of space and time, storage and transmission. Players play with these moments of hesitation, adjustment, fresh starts, and failures. These practices allow play to happen and to continue but also are themselves play acts; they are part of and constituent of play, especially play that, like mobile DS play, is shaped by and dependent on larger social, spatial, and temporal contexts that are in flux.

Portable play is contingent: the DS player reworks gaps in time, exploits spaces more or less open to play, and deals with other issues of context that can define and delimit their play activity as much as any code, game mechanics, or miniboss. Many of the ways in which players use the DS have to do with management of time and space in order to protect play or even just the possibility of play.

What we might get from a careful reworking of some of the issues of the relation between play and non-play and between player and plaything is something like Benjamin's "innervation," a kind of contact with, in, and through games, like the "dynamite of the tenth of a second" that he saw in cinema that could illuminate otherwise ignored or hidden aspects of our own lives.[24] The promise of cinema for Benjamin (only ever a promise) was that cinema could literally show us things that we could not otherwise see or that we would fail to notice, revealing to us something about our lives that would help us attain a new consciousness, even a revolutionary one. I am not suggesting that we will achieve such a conciseness through mobile gaming, nor am I arguing that such mobile gaming offers us the fantasy of escape via play but rather provides the possibility of illuminating the conditions of play in our current milieu, highlighting the way in which we deal with space and time that are not our own.

In exploring the details of how we share our mobile play or how we save games, and in pausing them, save play, we, scholars, critics, and players alike can learn something about the stakes and contexts for this medial situation we play and live in. We learn something about not just games, mobile or otherwise, when we examine the details of their management and usage. We can also learn about contingency and context by looking at how we play with the temporal and spatial dynamics of media. We can learn something about mobility, impermanence, social spaces, and the communicative possibilities of those spaces by playing with and thinking about playing with these kinds of portable systems. Reading play into save and pause functions connects these affordances to larger issues involving the management of space and time, and although making this connection does not fundamentally change what it is we do with the DS and mobile video game systems, it enables us to play with the constraints and contingencies of space and time.

Notes

1. Harold Innis, *The Bias of Communication* (Toronto: University of Toronto Press, 1964).
2. According to Nintendo's own estimation, 26.4% of players have used the online function: http://www.nintendo.co.jp/ir/en/library/historical_data/index.html
3. Bernard Suits, *The Grasshopper: Games, Life, Utopia* (Peterborough, ON: Broadview Press, 1978), 15.
4. James Newman, "The Myth of the Ergodic Videogame," *Game Studies: The International Journal of Computer Game Research* 2002 1, no. 2. Accessed August 13, 2010, www.gamestudies.org/0102/newman/.
5. Chantal de Gournay, "Pretense of Intimacy in France," in *Perpetual Contact: Mobile Communication, Private Talk, Public Performance*, ed. J. E. Katz and Mark Aakhus (Cambridge, UK: Cambridge University Press, 2002), 193–205.
6. Erving Goffman, *Encounters: Two Studies in the Sociology of Interaction* (Indianapolis and New York: Bobbs-Merrill, 1961).
7. Erving Goffman, *Encounters*, 21.
8. Walter Benjamin, *Reflections: Essays, Aphorisms, Autobiographical Writings* (New York: Schocken Books, 1986).
9. Walter Benjamin, *Reflections*, 79.
10. Jonathon Metts (2005), an amateur (and expert) game historian, breaks down the three ways the dual screens are used in DS games: a) the touch screen is not used at all; b) the touch screen is optional or is used for minor features; c) the touch screen is the primary method of control. See Jonathan Metts, "Can't Touch This," *Nintendo World Report*, March 3, 2005. Accessed June 15, 2011, http://www.nintendoworldreport.com/editorial/554.

11. The exception might be when one is about to play a stylus intensive game, such as Tecmo's *Ninja Gaiden Dragon Sword* (2008).

12. Nintendo EAD, *Legend of Zelda: The Phantom Hourglass* (Nintendo, 2007).

13. Indieszero, *Retro Game Challenge* (Xseed Games, 2009).

14. James Tobias, "Designing Wonder: Complexity Made Simple or the Wii-Mote's Galilean Edge," *Television & New Media* 11, no. 3 (2010): 197–219. James Tobias, in this article on Nintendo's Wii-mote, shows in a similar fashion how another Nintendo interface device can be treated both concretely as a gestural object in a material sense, as well as a conceptual-tool through which to rethink medial gestures and agency historically and critically as they relate to affective labor and wonder. In "Designing Wonder: Complexity Made Simple or the Wii-Mote's Galilean Edge," we see how the Wii and Wii-mote "support a specific intersection" of styles of play and hence an analysis of such styles and what is at stake in larger global of issues of "hyperindustry" (p. 213). Interface instruments like the DS stylus or the Wii-mote offer us useful points of contact between discrete player practices and "transformed effects, conditions and contexts on a larger scale" (p. 215).

15. Mark J. P. Wolf, "Time in the Video Game," in *The Medium of the Video Game*, ed. Mark J. P. Wolf (Austin: University of Texas Press, 2001), 77–93.

16. Mark J. P. Wolf, "Time in the Video Game," 83.

17. Harrison Gish, "Developing Transitional Space in Classic Games," (presentation at the annual meeting of the Southwest Texas Popular Culture and American Culture Association, Albuquerque, New Mexico, February 25–28, 2009).

18. Mark J. P. Wolf, "Time in the Video Game," 83.

19. Mark J. P. Wolf, "Time in the Video Game," 87.

20. Mark J. P. Wolf, "Time in the Video Game," 83.

21. "Temptation of completion increases, a diffuse subcutaneous malaise gnaws and festers to mobilize a new degree of caring for the first shot of the next attempt, and you play a bit better because each move is charged with an attentiveness reaching forward in anticipation." David Sudnow, *Pilgrim in the Microworld* (New York: Warner Books, 1983), 54.

22. Erving Goffman, *Encounters*, 26.

23. Again from Nintendo's sales estimates, it is worth noting that one unit likely equals more or less than one DS player: www.nintendo.co.jp/ir/en/library/historical_data/index.html

24. Walter Benjamin, *Illuminations: Essays and Reflections* (New York: Schocken Books, 1969).

You Can Ring My Bell and Tap My Phone

Mobile Music, the Ringtone Economy, and the Rise of Apps

BEN ASLINGER

On the November 30, 2005 episode of *CSI: New York,* Detective Danny Messer and his partner cross the street as his girlfriend calls him. Instead of a traditional ringtone, Messer's phone plays Coldplay's "Talk," a popular song about relationship dramas and miscommunications. After hanging up and entering the crime scene, the detectives make fruitless attempts to find clues, until Messer says they need to use a new light. The blue light they subsequently employ reveals walls full of writing as Coldplay's "Talk" plays again, making the point through the non-diegetic track that Messer and his partner have used technology to make the crime scene "talk" to them in a way impossible with the naked eye. The fact that the ringtone was something that a U.S. broadcast television series could mine for both promotional and narrative effect testifies to the ringtone's visibility as an industrial and cultural phenomenon.

Music labels and publishers seemed ecstatic about the possibility that ringtones could provide new revenue streams and revitalize the global music market as CD revenues plummeted, piracy rates skyrocketed, and original track sales from online stores such as iTunes failed to grow at the rates desired by executives in music and technology firms in the early to mid-2000s. This essay examines the hype surrounding the emergence of the ringtone market, changes in the ringtone and mobile music markets that affect which genres and musical cultures are dominant, and the shift from a ringtone to an app economy where there is less emphasis on selling iso-lated tracks and more emphasis on selling user experiences which are connected to musical works or performer identities. This essay adds to a scholarly conversation on ringtones and mobile music begun by writers such as Heikki Uimonen, Imar de Vries and Isabella van Elferen, Gerard Goggin, and Christian Licoppe.[1]

While this essay is not intended to be an exhaustive review of all things writ-ten about the ringtone and mobile music, I draw from a close reading of over 1,800 popular press and trade press stories on ringtones and mobile music from 2000–present. I begin by sketching out a brief history of the rise and fall of the ring-tone. I then discuss how the ringtone produced public anxieties about youth cul-tures, noise, and consumer protection. Finally, I briefly describe the emerging app economy.

I have tried to include data and information on ringtones and mobile music from a variety of national contexts. This is extremely important given the ways that services such as i-mode from Japan's NTT DoCoMo helped popularize new ways of engaging with media via mobile devices and how ringtone economies developed in Europe and East Asia years before ringtones were available, much less profitable, in North America. However, it is difficult to paint a complete picture of the mobile music soundscape because of the following factors: struggles over how to define the music industry;[2] struggles over power within the recording and music industries between artists, publishers, performance rights organizations, major labels, and "indie" labels; the emergence and decline of aggregators; the number of wireless car-riers operating at the multinational, regional, national, or local level (with different reporting mechanisms and systems for billing mobile media content); and conflicts between carriers, aggregators, labels, and publishers over who controls the mobile music market, modes of distribution, and who "owns" the consumer.

The Evolution of the Ringtone Market

The only new media distribution outlet that excited the music industry from the outset was the mobile phone. Though revenues from mobile games, television, and video excite media executives today, the development of the ringtone economy and carriers' development of music stores were the first steps in exploiting wireless capabilities in delivering media content.

In 1997, Vesa-Matti Paananen, a Finnish programmer, developed Harmonium, a software program that enabled users to program phones to make and process musi-cal sequences they could send to friends.[3] In line with many developers' enthusi-asm with open source software and excitement surrounding the Linux operating system, Paananen made Harmonium free. Quickly, however, ringtones transcended from a form of freeware to a commodity that required partnerships between firms interested in exploiting and controlling the ringtone. Interest in the global potential of the cell phone ringtones and wireless music resulted in a growing number of partnerships between music labels, manufacturers, and carriers. From 2002 to 2006, handset manufacturers such as Motorola, Nokia, and Ericsson debuted new multifunctional handsets and rolled out new business initiatives in content provision that would later anger wireless carriers. Emerging partnerships between wireless carriers and major record labels and content providers included Vodafone's agreements with Warner and Sony, Universal Mobile's deal with French telecom group Alcatel, Warner's pledge of cooperation with Deutsche Telecom's T-Mobile, Virgin Mobile USA and Universal Music Group's commitment to work together, and MTV Networks' decision to distribute content through Verizon Wireless' network.[4] Content providers began to experiment with offering ringtones directly to consumers, and News Corporation paid $188 million for controlling interest in VeriSign's Jamba! Ringtones unit in a move designed to strengthen its cell phone entertainment company Mobizzo.[5] Sprint Nextel unveiled a mobile music store, and Verizon launched its VCast service;[6] other carriers followed suit, attempting to take a larger piece of the ringtone revenue pie and channeling users to carrier-operated music stores.[7] By the middle of the decade, the ringtone was seen as such a profitable revenue stream for carriers and labels that the number of global partnerships seems impossible to count. At the same time, the rapid increase in the number and types of stakeholders invested in the ringtone's profitability resulted in conflicts over who controlled the mobile music market.

Ringtones were seen as a way to reinvigorate global sales of music. Internationally, the polyphonic ringtone market was worth $3.1 billion in 2004.[8] In 2004, Korean carrier SK Telekom's use of ringback tones (heard by the caller while one is waiting for the receiver to pick up) helped the carrier achieve 30 percent market penetration.[9] In 2005, the size of Japanese market was estimated to be 12 billion yen ($113 million) annually, with most revenue coming from "chaku-uta" or master ringtones.[10] By the end of 2005, ringtones accounted for more than one-tenth of the global music market and were the most profitable non-voice service.[11] Illustrating the growing importance of the mobile phone in music commerce, two-thirds of all Italian music downloads in 2006 came from cellphones.[12] For carriers, ringtones were a new revenue stream, especially in markets where a large percentage of the population already had a mobile phone. For the music industry, ringtones were a way to take advantage of (rather than lose to) the Internet networks.

In the U.S. in 2004, ringtone revenues were estimated at $300 million.[13] In April 2005, 24.6 million people or 13.6 percent of American cell phone users downloaded a ringtone.[14] By 2008, however, the boom years of the ringtone were over. Warner Music Group announced in its first quarter report that mobile sources accounted for only 30 percent of digital revenue; revenue from mobile music decreased by 17.6 percent from 2007.[15] The decline of the ringtone forced the music industry to think about how to reinvent mobile music commerce for a new generation of 3G-enabled smartphones.

Evidence suggesting that ringtones and song downloads were popular came from the *Billboard* ringtones charts and sales figures, but mounting frustrations surrounding pricing and access suggest that the romance surrounding wireless music commerce was short-lived. In July 2005, Dwight Silverman called master ringtones "greedtones," given that online music stores such as eMusic and Apple's iTunes offer cheaper, better quality downloads.[16] Consumers wondered if the immediacy of over the air downloads to handsets was worth 2 ½ times the price of a legal download from iTunes. In November 2005, Walter Mossberg argued that the "bumbling record industry" and wireless carriers, whom he dubbed the "Soviet ministries," were responsible for the high prices for wireless services.[17] Mossberg's anger at the "Soviet ministries" stemmed from their refusal to allow open competition for wireless services and their attempts to restrict the ways consumers can access music. The major U.S. wireless carriers have preferred a "walled garden" approach—a term used to designate the proprietary screen/"deck"/storefront that Sprint and Verizon have preferred. This has made off-portal sales (e.g., using one's Sprint phone to buy ringtones and music from other places besides the Sprint Music Store) and the sale of ringtones through short codes (traditionally numbers sent via text message) less commercially successful in the U.S. market. By 2005, only 10 percent of U.S. wireless music transactions were off-portal sales or sales stemming from short codes; in stark contrast, off-portal sales accounted for over 80 percent of the wireless music market in Europe.[18] The use of short codes sent via SMS to purchase ringtones challenges the Americanization thesis in two ways, since SMS became popular in the U.S. years after it had become ubiquitous in Asia and Europe and ringtone economies were built in the U.S. after ringtones had become profitable elsewhere.

Wireless and music firms were most enthusiastic about the ringtone's potential with niche demographics. Young adults were the first to be targeted as ringtone consumers. John Burris, Sprint's director of wireless data services, said in 2005 that teens were the "sweet spot."[19] A study by mobile media measurement group Telephia in November 2005 found that women bought 69 percent of ringtones.[20] Forrester Research found that "20% of African-American mobile subscribers regularly use mobile data services and download content, compared with only 7% of whites."[21] Nielsen Interactive Entertainment estimated that "69% of Hispanics, compared with 90% of African-Americans, have either downloaded a ringtone or are interested in custom ringtones, MP3 players, streaming multimedia or FM radio."[22] While ringtones were seen as a way for the music and wireless industries to target niche demographics, the market utopianism and irrational exuberance surrounding the ringtone were soon supplanted by the view that ringtones had been a profitable, but short-lived fad. All parties concerned began rethinking the nature of the mobile music marketplace.

This Little Ringtone Went to Market

As the music industry saw the potential profits of cell phone ringtones and music stores, it attempted to dominate negotiations with cell phone handset manufacturers, wireless carriers, billing agents, and content aggregators. Artists and labels had to decide when and why to sign exclusive and non-exclusive deals with wireless operators.[23]

At first, negotiations surrounding licensing rates and revenue sharing took place between four major players: wireless carriers, billing agents, content aggregators, and the music industry (labels, publishers, and/or performance rights organizations). Carriers sometimes handle billing in-house; however, they sometimes choose to subcontract this to billing agents such as Qualcomm or QPass. Content aggregators denotes a class of businesses who package media content for mobile devices and work as the liaison between carriers who want to exploit musical content and the music industry which controls the conditions under which music flows into new delivery technologies. The largest American content aggregators have been Zingy, Faith West, and Moviso; once independent, they now belong to transnational firms with interests in wireless telephony. Many content aggregators were bought up by larger wireless firms and those that remained independent were increasingly squeezed by carriers and the major labels.

Technical advancements in handset sound quality along with increases in cell phone storage space changed the types of ringtones available. First generation ringtones were monophonic ringtones; these ringtones were played one note at a time (think basic ringtones such as the bell, whistle, alarm, or the drum line of a popular song). When Wireless Access Protocol-enabled cell phones with better sound quality came on the market, content aggregators became more interested in providing polyphonic ringtones. A technical advance over monophonic tones, polyphonic ringtones could feature multiple notes/instruments at a time. Polyphonic ringtones were synthesized segments of songs that only required aggregators to deal with music publishers and/or performance rights organizations, since no sampling of original recordings was involved. When more advanced phones came on the market, the music industry became interested in master ringtones. Master ringtones, also called "truetones," were segments of the original sound recording; selling them required content aggregators to work with record labels in addition to publishers.

Wireless carriers, who handle billing and collection, keep between 10 and 40 percent of gross music revenues; shares are higher if the carrier hosts the ringtones and does in-house billing verification. Carriers have historically insisted that they deserve a hefty percentage since they have had to upgrade their infrastructures, improve reception, and address bandwidth concerns. If third-party billing verification firms such as Qualcomm or QPass get involved, these firms get 10 to 16 percent of the revenue. After this, content aggregators claim their percentage. In the case of polyphonic ringtones, aggregators earned between 35 and 65 percent of the music revenue—out of which they paid publishers or performance rights organizations. However, when master ringtones came on the market, aggregators had to deal with the major record labels, who asked for around 50 percent of retail. This lowered aggregators' shares to 15 to 20 percent; in some cases, it erased any profits.[24]

The squeezing of content aggregators was important because these firms worked to understand consumer desires and as liaisons between the music industry and wireless carriers.[25] Aggregators had done their homework, working to provide hip hop polyphonic and master ringtones that would perform well in the teen-driven wireless music market. But aggregators had also helped independent labels and artists get a shot at a piece of the ringtone pie. In 2005, Vodafone head of music Ed Kershaw said, "For an independent label wanting to get involved in ringtones and realtones, which are still worth far more than full-track downloads in revenue terms, I think their only choice is either to do a really good deal with a content creation company or go down the aggregator route."[26] Indies had to deal with specialist aggregators, as carriers would only do business with major labels.[27] With aggregators squeezed or put out of business, major labels and major carriers assumed dominant control over the mobile music market.

Mobile Music Genres and Musical Quality

Hip hop worked well for wireless distribution because the genre worked to target core demographics and worked well within early 2000s handsets' sonic constraints. Hip hop's generic conventions—a reliance on bass and percussion, sampling, and repetition—made it possible for aggregators to digitally synthesize a section of a song and have the segment recognizable as a rendition of a popular tune. In 2003, when Beyoncé's "Baby Boy" and "Crazy in Love" sold over 500,000 units and raked in over $1.25 million in revenue on the Sprint PCS network in July, these successes illustrated the commercial potential of master ringtones.[28] The first week of November 2004 illustrated the commercial importance of hip hop ringtones even further, when Usher and Alicia Keys' "My Boo" sold over 97,000 units; the top legally downloaded song, U2's "Vertigo," only sold 25,000 units.[29]

Cell phones capable of processing the master ringtone, however, could also produce sounds clear enough to make rock, country, and folk music viable wireless music markets. Sasha Frere-Jones writes, "Musical genres that suffered as polyphonics—sonically thick guitar rock, country, and jazz—can now challenge the hip hop hegemony."[30] Rock, country, and folk tunes proved popular in trying to court other generational demographics. Incorporating these genres into wireless music commerce allowed the industry to target consumers over the age of 30 and professional/business consumers wanting to personalize their BlackBerries and other "smart" phones.

Ringtones also prompted debates about musical quality and control. Artists such as Abba wished ringtones to be made from master recordings and not from synthesized reproductions of songs, leading some artists to try to refuse permission for ringtones and other forms of "new" media distribution they feel would tarnish their image or degrade their music; however, control over the distribution of polyphonic ringtones was difficult, if not impossible, to enforce.[31] Tony Dimitriades, artist manager, says, "There is a sense among some that it [the ringtone] bastardizes the music, takes away the sincerity and the original intent of the artist."[32] A 2007 dispute between Bob Marley's family, Verizon Wireless, and Universal Music raised the issue of whether artist permissions were legally necessary and how much bad PR would be generated by creating ringtones without the artist's or the artist's estates permission.[33]

The desire to distinguish between polyphonic and master ringtones led to legal definitions of the truetone in the U.K., where "companies will only be able to use the phrase 'realtone' if they are selling a ringtone produced from an original recording 'with the performer clearly identified in its marketing material.'"[34] Graeme Samuel, chairman of the Australian Competition and Consumer Commission, indicated that "ringtone promoters have been asked to replace the words 'real' or 'true' tones with the term 'cover tones' to better reflect the ringtones being sold."[35]

Mobile Music, Public Anxieties

When basic monophonic ringtones were the norm, ringtones emerged as a popular way to personalize your handset and as a simple tool for identifying when you were receiving a call. Grant Edwards writes, "Ever notice the number of people that dive into pockets and handbags or reach for their belts when a mobile rings in a crowded room? Hence the need for more ringtones."[36] When Edwards wrote these words in 2001, ringtones were perceived as a way to avoid public fumbling in one's pockets, purse, or satchel—an auditory cue that identified a clear handset-user match. To Edwards, more ringtones meant more opportunities for individualization; clear handset-user matches meant that one could ignore phone rings with confidence.

Very quickly, however, the idea of ringtone as signal—as meaningful sound designed to communicate auditory cues to the mobile phone owner—was replaced with the idea of ringtone as noise—unwanted, undesired, and irritating sonic intrusions into the physical environment. Complaints about the publicness of the ringtone can be seen in the following accounts:

> History records that when the first phone call was made on March 10, 1876, the first words spoken were: "Come here Watson, I want you." What is not recorded, is whether the world's first phone blasted out the latest hit from the music hall before Alexander Graham Bell's assistant picked it up. Once there was just a plain bell ring. Then a cheeky chirruping….Now every mobile comes with a cacophony of tinny music from the charts and television.[37]

> During an average morning at work I'm assaulted by Muppets, irritated by The Simpsons, infected by Kylie, and even made to cringe by Celtic and Rangers anthems. In any shop the chances are the audience senses will be invaded by James Bond, Will from Popstars or even a classical touch of Tchaikovsky.[38]

> All passengers share the same space. Every passenger who pays to ride on a train expects to enjoy a reasonably pleasant ride, and not be assaulted by loud ringtones and talk. [39]

> On the bus, in the office, at the pub—you cannot escape the cacophony of the mobile phone ringtone. [40]

What is striking is the strength of the language, with words such as infected and assaulted being bandied about casually by journalists and the judgment that the ringtone is decidedly unmusical. The ringtone is judged not only to be noise, but to be the worst possible kind of noise—a cacophony from which there is no hope of extracting a meaningful signal.[41]

While ringtones became increasingly seen as audible nuisances and irritants, the marketing and sale of ringtones brought mobile media commerce under greater public scrutiny from governments and consumer watchdog organizations. In 2005, the Japanese Fair Trade Commission ruled that Sony Music Entertainment, Avex, Universal Music K.K. Victor Entertainment, and Toshiba-EMI "violated the unfair-trading-practices section of the Anti-Monopoly Law by consigning the provision of master ringtones exclusively to Label Mobile, a Tokyo-based company they jointly own."[42] While performance rights organizations such as ASCAP (American Society for Composers, Authors, and Publishers) believed that ringtones should be counted as public performances, U.S. District Judge Denise Cote ruled that these downloads were not performances.[43] The distinction between download and performance affects the operations and future legal status of performance rights organizations such as ASCAP, BMI (Broadcast Music Inc.), and the recently created Sound Exchange. Cote's ruling also highlights a disconnect between commonsense and legal definitions of publicness.

Most of the public anxiety, however, centered on what many saw as unethical and misleading appeals to the youth market. Journalist Fergus Sheppard writes, "To children, teenagers and a few others, ringtones are, de facto, the new rock'n'roll."[44] Rock'n'roll provoked moral panics over taste, musicianship, and youth culture, and as products like the Crazy Frog ringtone began to outsell traditional bands like Coldplay, ringtones became a part of the debate on the susceptibility of youth and the media youth were consuming. Toby Walne writes, "Mobile phone ringtones were once just a minor irritant on public transport. Today, they're a lucrative multi-million-pound industry with member companies investigated by watchdogs over price, sales transparency and marketing to minors who are unaware of the costs."[45] *Billboard*'s Antony Bruno reported in 2008,

> Most off-deck content providers operate subscription services whereby members can download a set number of ringtones, wallpapers and games for a monthly fee, typically $10. They attract new customers by advertising free or deeply discounted ringtones, many of whom don't notice the subsequent extra recurring fee....But recent developments put wireless operators in the cross hairs for their involvement in collecting the disputed charges, even though they play no role in marketing or distributing the content in question. [46]

Anxieties over breaches of consumer protection rules led EU consumer affairs commissioner Meglena Kuneva to promise action to curtail ringtone scams.[47] Carriers were implicated because they collect and bill charges. But the ways that wireless carriers used consumer protection concerns to argue for greater control over mobile music are eerily reminiscent of the broadcast networks' rhetoric used in the 1950s to wrest control of broadcasting from advertisers. Recording labels also had vested interests in getting rid of the bad apples and delegitimizing aggregators and independent firms in order to fight disintermediation.

Anxieties about ringtones and youth were similar to previous moral panics surrounding rock, punk, and techno not only in terms of the "susceptibility" of youth audiences, but also in terms of debates about taste and musical quality. Many critics were dismayed that adolescents preferred ringtones to full-song downloads.

Consumer protection worries along with the fact that teens bought ringtones at all worked to construct teens as naïve listeners. Angela Pacienza writes, "A funny thing happened this year in the digital music universe—teens were inspired to pry open their Hello Kitty wallets for 30-second ringtones while four-minute song downloads still couldn't shake loose a single penny."[48] Headlines such as "Ringtone leapfrogs Coldplay to top charts" highlight the ringtone's commercial rise as a competitor and a threat to conventional musicianship.[49] The stereotype of the unsavvy teen led critics as well as the wireless and music industries away from more nuanced discussions about the generational relevance of music and how people determine what is valuable enough to pay for. Critics (both popular and scholarly) need to highlight how the nature of musicianship alters with new technologies for distributing and creating music.

The Shift to Apps

Rising from low monophonic sales, increasing with the rise of polyphonics, and then cresting with realtones in the mid-2000s, ringtones were the first major step in crafting a distinct mobile music culture. There were signs that the music industry recognized the short shelf life of the ringtone. Rio Caraeff, general manager of Universal Music Mobile, told *Billboard* in 2007, "If the ringtone market falls apart…[ellipses in original] we're obviously at more risk because we're not as diversified as I'd like to be."[50] Antony Bruno wrote in 2008, "Mobile is no more a panacea than online downloads or subscription services."[51]

Ringtones set a disturbing precedent for mobile media commerce with limited personalization and circumscribed interactivity; journalist Michael Tedeschi notes that "not all songs and artists are licensed for downloads and you don't get to pick which 30-second segment of the song is played."[52] Mobile and networked music services such as Pandora and Spotify must deal with an industry that loved the ringtone precisely because of its interactive veneer; new music startups that are actually interactive pay through the nose and often limit interactivity (e.g., in Pandora's limitation on the number of songs a user can skip) in order to be able to pay license fees that are not crippling (although still a huge portion of a firm's operating budget).

The fall of the ringtone has forced the recording and music industries to reconsider what mobile music commerce is. In 2009, *Billboard's* Antony Bruno wrote:

> To labels, a ringtone is a music sale. But to mobile users, the ringtone is an application they use to personalize their phone . . .
> The mobile phone isn't a point of retail, but a point of access. . . .
> As mobile operators focus more on the sale of data access than on multimedia downloads and entertainment subscriptions, the music industry can expect a further

decline in its revenue from mobile content sales. Other than an increase in streaming royalties and a possible uptick in mobile track sales through music-based apps like Pandora and Shazam, this shift to an app-based mobile content economy won't benefit labels, music publishers or artists unless they revamp how they're compensated.[53]

Given the confusion about what the ringtone was and its value, it is not surprising that negotiations for licensing streaming services and mobile apps are so fraught with complexity; the music industry is in still in the midst of rethinking mobile strategy.

In 2005, Eric Winbolt, digital media manager at EMI Records, said that mobile media needed to "become a place where you can nurture and develop talent, not just another revenue stream from artists."[54] However, ringtones were seen as a hit- and chart-driven phenomenon that privileged major label and A-list artists; they were never meaningful ways of engaging with fans or ways of introducing new artists. At the most, ringtones were an opportunity for artists to cut their teeth on mobile media before moving on to do "something unique."[55]

In 2006, Tom Ryan, senior VP of digital and mobile strategy at EMI described his vision of the "mobile album." He said, "Fans want to buy content based on the artist, not based on a specific product. Just like a CD is a bundle of tracks, a mobile album could be a bundle of mobile products around one artist."[56] Ryan's hopes for a reinvention of the album may be fulfilled with the launch of Björk's *Biophilia*. Debuted as a series of multimedia live performances at the Manchester International Festival in June and July 2011, Björk plans a series of "companion apps" that will be central to the *Biophilia* project.[57] While imaginative apps from artists such as the U.K. group The XX, the U.S. rapper T-Pain, The Streets, and David Bowie have garnered critical attention, many music apps are unimaginative ways to bleed money from fans or are simply repetitions of content that can be easily accessed for free via artists' websites.[58] Salmon writes, "Musician-related apps are only going to get more ubiquitous. Not just because they still seem to lend an act an air of cutting-edge cool, but because—at a time when many listeners simply can't comprehend why you would pay for music—people will happily shell out a few quid for an app."[59] As apps for artists become more ubiquitous, there will be increased pressure to create unique and engaging content so that this portion of an artist's promotional and content strategies will be worth the time and investment.

More exciting and more visible are apps such as Spotify, Pandora, and Shazam. A 2010 report co-authored by representatives from the Pew Internet and American Life Project and Nielsen found that 43 percent of Nielsen downloaders had used music apps in the past month, Hispanics were the ethnic group most likely to have used a music app in the recent past, and "music apps ranked second on the most downloaded list" and "fifth on the most used list."[60] Music apps, in order of popularity, were Pandora, YouTube, iPod/iTunes, Shazam, and Yahoo Music.[61] Music apps were also found to be the second most common paid app, behind games—10 percent of the sample had paid for a music app.[62]

A study by MIDEM and Nielsen found that over 20 percent of respondents has streamed music through a mobile phone from an app and that around 20 percent had downloaded/used music apps on a mobile phone; however, in 81 percent of the cases, music apps had been downloaded for free.[63] MIDEM and Nielsen found that music discovery apps, artist apps, and streaming apps were the most popular, with artist apps doing best in Europe and music discovery apps (including Shazam, Rhapsody, Napster, and Pandora) doing best in North America and the Asia Pacific region.[64] The report asks an intriguing question that it does not (and could not) answer: "Are music apps considered as a 'nice to have' by consumers, or are they a great marketing tool for the industry, or even a new source of significant revenue?"[65] Debates over the licensing of music for mobile media (especially for streaming services) and the tug of war in the app marketplace between the open ecosystem of the Android marketplace versus the tightly controlled ecosystem of the Apple App Store will undoubtedly affect the answer to the above question.

Conclusion

The ringtone's articulation with youth culture, connection to anxieties about the publicness of music, the declining frequency of the phone call as a pattern of mobile use, the increasingly crowded ringtone marketplace, the high price of ringtones, and the proliferation of free ringtone sites meant that ringtones (if they ever lived up to the hype surrounding them in the music industry trade press) could only have dominated the imaginations of music executives and consumers for a brief time. Music on the mobile phone has shifted from monophonic to polyphonic to master ringtones, from ringtones to app, and from user customization to interactive experiences. Mobile music has the potential to reconfigure how we come to discover music and alter how music fits into our daily habits, routines, and travels. However, only time will tell how the tensions between art and commerce will change listening practices as well as the shape of music in the data stream of mobile media.

Notes

1. Heikki Uimonen, "'Sorry, Can't Hear You! I'm on a Train!' Ringing Tones, Meanings and the Finnish Soundscape," *Popular Music* 23, no. 1 (2004): 51–62; Imar de Vries and Isabella van Elferen, "The Musical *Madeleine:* Communication, Performance, and Identity in Musical Ringtones," *Popular Music and Society* 33, no. 1 (2010): 61–74; Gerard Goggin,

"Mobile Music: Ringtones, Music Players, and the Sound of Everything," in *Global Mobile Media* (New York: Routledge, 2011), 55–79; Christian Licoppe, "The Mobile Phone's Ring," in *Handbook of Mobile Communication Studies*, ed. James E. Katz (Cambridge, MA: MIT Press, 2008); James E. Katz, Katie M. Lever, and Yi-Fan Chen, "Mobile Music as Environmental Control and Prosocial Entertainment," in *Handbook of Mobile Communication Studies*, ed. James E. Katz (Cambridge, MA: MIT Press, 2008).

2. Dave Laing, "World Music and the Global Music Industry: Flows, Corporations, and Networks," *Popular Music History* 3, no. 3 (2008): 212–231; John Williamson and Martin Cloonan, "Rethinking the Music Industry," *Popular Music* 26, no. 2 (2007): 305–322.

3. Sasha Frere-Jones, "Ring My Bell: The Expensive Pleasures of the Ringtone," *The New Yorker*, March 7, 2005, 86.

4. Ben Fritz, "Warner Calling," *Video Business*, April 12, 2004, 10; Juliana Korantang, "Gold Rush Is on in Mobile-Music Sector," *Billboard*.com, June 26, 2004; "Hollywood Records, Xingtone Team on MP3 Ringtones," *Business and Industry Online Reporter*, January 31, 2004; "Cell Phone Users Want Mobile Music," *Business and Industry Online Reporter*, July 17, 2004; "Ringtone Album Sees Better-than-Expected Sales," *Business and Industry Online Reporter*, August 21, 2004; "Sony Delivers Mobile Music Service," *Business and Industry Online Reporter*, June 19, 2004; "Virgin Mobile Scores UMG Exclusive," *Business and Industry Online Reporter*, June 12, 2004.

5. Harry Maurer, "Pricey Ring Tones," *Business Week*, September 25, 2006, 35.

6. "Nearly 14% of US Mobile Users Bought a Ringtone in April," *Business and Industry Online Reporter*, June 4, 2005.

7. Terence Keegan, "The Mobile Media Network; Carriers: Music on Cell Phones Is Just the Beginning," *Medialine*, March 1, 2006, 24.

8. Jim Farber, "Ringtones Outsell Legit Song Downloads 3 to 1," *The Gazette* (Montreal), November 6, 2004.

9. Scott Banerjee, "Ringback to the Future," *Billboard*.com, November 6, 2004.

10. Steve McClure, "Japan: Labels May See Court," *Billboard*.com, April 16, 2005.

11. Himanshu Rai, "Lord of the Ringggz. .zo," *New Straits Times* (Kuala Lumpur), October 2, 2005.

12. Yuki Noguchi, "Ringing Up Big Music Sales; Cellphones Supplant Radio in Promotions," *TechNews*, May 13, 2006.

13. Scott Banerjee, "Ringback to the Future," *Billboard*.com, November 6, 2004.

14. "Nearly 14% of US Mobile Users…," *Business and Industry Online Reporter*, June 4, 2005.

15. "Bits and Briefs: Ringtone Reduction," *Billboard*.com, May 24, 2008.

16. Dwight Silverman, "Pay Just Once for Software to Get Unlimited Ringtones," *Houston Chronicle*, July 5, 2005.

17. Walter Mossberg, "Sprint Brings Tunes to Your Phone—for a Price," *Chicago Sun-Times*, November 19, 2005.

18. Ibid.

19. Jyoti Thottam, "How Kids Set the (Ring) Tone; In a Wireless World, Teenagers are Driving the Hottest New Technologies Since the Dotcom Era," *Time*, April 4, 2005, 40.

20. "The Latest News from the .Biz," *Billboard*.com, November 19, 2005.

21. Antony Bruno, "Def Jam Hears Call for Wireless Content," *Billboard*.com, June 11, 2005.

22. Leila Cobo, "Latin Novas: Sivam Scholarships Keep Mexico Musical," *Billboard*.com, April 8, 2006.

23. Antony Bruno, "AT&T for Two," *Billboard*.com, December 15, 2007.

24. Scott Banerjee, "Ringtone Rumble Brewing," *Billboard*.com, May 22, 2004.

25. Juliana Korantang, "Gold Rush Is on in Mobile-Music Sector."

26. "The Specialist Role of the Aggregators," *Music Week,* September 24, 2005.
27. "How to Get Your Music Mobile: Foreword," *Music Week,* June 25, 2005, 52.
28. Scott Banerjee, "Ringtone Rumble Brewing."
29. Jim Farber, "Lords of the Ring: Alicia, Usher Top New Phone Tone List," *Daily News* (New York), November 2, 2004.
30. Sasha Frere-Jones, "Ring My Bell," 86.
31. Adam Sherwin, "It's Abba on the Phone Making a Lot More Money, Money, Money," *The Times* (London), April 19 2006.
32. Jeff Leeds, "Record Companies Chase the Profit from Mobile Ringtones," *Sydney Morning Herald,* August 19, 2004.
33. James Doran, "Marley Family Rise Up against Ringtones Deal," *The Observer* (London), September. 23, 2007.
34. David Derbyshire, "Ringtone Guide to End Soundalikes Confusion," *The Daily Telegraph* (London), January 24, 2006.
35. Anthony Keane, "Caught Ringing Off-tone," *The Advertiser* (Australia), October 16, 2006.
36. Grant Edwards, "Download It Again, Sam," *The Advertiser,* July 14, 2001.
37. Alun Palmer, "Ringtone Rage," *The Mirror,* March 22, 2004.
38. Brian Beacom, "Can't Get You Out of My Head," *Evening Times* (Glasgow), May 18, 2002.
39. "Schoolbag Burden," *The Straits Times* (Singapore), October 29, 2006.
40. Kathy McCabe, "Incoming Discord of These Lowered Tones," *The Daily Telegraph* (Sydney), July 17, 2004.
41. I am indebted to Kate Crawford's discussion of signal and noise in mobile and social media, elaborated in her Berkman Center for Internet and Society talk that can viewed and downloaded at http://cyber.law.harvard.edu/interactive/events/luncheon/2010/08/crawford.
42. Steve McClure, "Japan: Labels May See Court," *Billboard*.com, April 16, 2005.
43. Jacqui Cheng, Judge: Ringtones Aren't Performances, So No Royalties," *Ars Technica,* October 15, 2009, http://www.arstechnica.com/tech-policy/news/2009/10/judge-ring-tones-arent-performances-so-no-royalties.ars?utm_source=rss&utm_medium=rss&utm_campaign=rss.
44. Fergus Sheppard, "Hopping Mad," *The Scotsman,* May 26, 2005.
45. Toby Walne, "Warning Bells over Mobile Ringtones," *The Independent* (London), July 10, 2005.
46. Antony Bruno, "Off-Deck and Out of Control?," *Billboard*.com, June 7, 2008.
47. Paul Cullen, "Commission May Close Sites Offering Ringtones," *The Irish Times* (Dublin), July 18, 2008.
48. Angela Pacienza, "Teens Insist on Free Downloads," *The Toronto Star,* December 26, 2004, G18.
49. "Ringtone Leapfrogs Coldplay to Top Charts," *The Herald* (Glasgow), May 30, 2005, News, 3.
50. "Special Feature: 2007 Top Mobile Executives," *Billboard*.com, October 27, 2007.
51. Antony Bruno, "Learning to Love," *Billboard*.com, September 13, 2008.
52. Michael Tedeschi, "Customized Bells and Whistles for Cellphones," *The Washington Post,* October 15, 2006.
53. Antony Bruno, "Change in the Air," *Billboard*.com, August 15, 2009.
54. "Wireless: Musical Movements," *New Media Age,* August 4, 2005, 23.
55. Antony Bruno, "Latin Fans? It's Mobile on the Line," *Billboard*.com, September 3, 2005.
56. Antony Bruno, "Disconnected," *Billboard*.com, September 16, 2006.
57. Chris Salmon, "Do Björk's Apps Break New Ground?," *The Guardian* (Music Blog),

March 17, 2011, http://www.guardian.co.uk/music/musicblog/2011/mar/17/manchester-international-festival-bjork-apps.

58. Ibid.; Sean Michaels, "David Bowie to Release Golden Years iPhone App," *The Guardian* (Apps Blog), April 5, 2011, http://www.guardian.co.uk/music/2011/apr/05/david-bowie-golden-years-iphone-app.
59. Chris Salmon, "Do Björk's Apps Break New Ground?"
60. Kristen Purcell, Roger Entner, and Nichole Henderson, *The Rise of Apps Culture* (Washington, D.C.: Pew Research Center's Internet and American Life Project, 2010), 5, 7.
61. Ibid., 28.
62. Ibid., 33.
63. Nielsen and MIDEM, *The Hyper-Fragmented World of Music: Marketing Considerations and Revenue Maximisation,* March 2011, 5, 16.
64. Ibid., 16–17.
65. Ibid., 16.

Appropriation of Cell Phones by Kurds

The Social Practice of Struggle for Political Identities in Turkey

BURÇE ÇELIK

Introduction

This essay seeks to explore the role of cell phones within the political and cultural struggle of young Kurds in Turkey, a country where Kurdish citizens, who make up more than 10 percent of the total population, have been denied their cultural rights. In particular, it focuses on how the cell phone, through its containing space for textual, visual and sonic content, is integrated into the everyday politics of young adult Kurds as a means of claiming their cultural identity, asserting their mother language as a language of technology and modernity, and producing a ritualized experience of everyday life where the medial Kurdishness is fully played out.

Recent research on cellular telephony reveal that this technology of "connected presence"[1] and of display is especially appealing amongst young people across different socio-technical contexts.[2] The instrumental value of the cell phone, such as the way it functions for the coordination of everyday life practices and communication and sharing of information and its expressive value, displaying the users' taste, identity and lifestyle, have been well covered in the literature on cellular telephony.[3] Recent studies have also shown that the cell phone has been domesticated in places whose cultural, economic and political relations to globality and Western modernity is characterized by asymmetries and has obtained distinct meanings and functions in those places.[4] While the link between the collective desire for modernity and globality and the passionate use of cell phones in these places is well established, the political use of this technology in these spaces as an imagined agent of equality, justice and political change has received limited scholarly attention.[5] Some works have drawn attention to the necessary problematization of youth as a homogenous concept and have argued for the diversity and difference in the relationship between young people and communication technologies based on ethnicity, gender and social class.[6] However, the literature is still short of studies that contribute to exploration of the ways in which ethnic and political youth appropriate the cell phone for their ethno-political purposes in less privileged areas, where the expressions of particular ethnic and political identities are severely limited. Drawing upon empirical data from a research project in Diyarbakır (the largest Kurdish-populated city in Turkey) and İstanbul (the largest city in Turkey), this work seeks to contribute to this field by showing how the cell phone finds its own use in the hands of young adult Kurds, to assert their cultural and political identity and to transform the whole communication environment, including sonic, visual and textual experiences, to embrace Kurdish identity, language, music and visuals.

My analysis of cell phones in Kurdish-populated areas of Turkey is grounded in the argument that things, tools or machines become technologies as they respond to users' wills, purposes, inclinations, interests, responses and desires. In this regard, "a technology is always at any given moment socially located. It is always implicated in social struggle."[7] The desired ends of the struggle and the way the struggle is carried out are inevitably bound to the socio-political and economic conditions that give rise to a particular domain of use of a technology. For Turkey's young adult Kurds, the historical, social, political, and cultural struggle is to garner an official acknowledgment of their existence by Turkish authorities as a unique cultural group with a distinct languages, history and memory, as well as to have warranties allowing the free exercise of their cultural identity, including the use of their mother language in private and public spaces. I consider the cell phone as a "container"[8] of not only visual, sonic and textual data or functions that enable daily communication, such as SMS, telephonic interaction and Internet connection, but also of technological, historical, ideological, political, and social projects produced by the actual users.

The social practice of cellular telephony is shaped by the practices of texting and talking, the pirate music industry, sharing of visuals, historical resentment, desire for political change, and love for the imagined practices of Kurdish culture and language. The cell phone, whose most distinctive characteristic is its proximity to the body of the user,[9] is also a gate that opens to the street where much of the political and everyday culture and resistances are manifested and reproduced. In this respect, this work uses the cell phone as a "tool" for understanding how technology, ideology and individual and collective struggles come together to produce an everyday practice and imagination through mobile technologies which, assumedly, can overthrow the dominant language and culture of the oppressed areas.

Contextualizing the Kurdish Issue of Turkey

The Turkish Republic was founded in 1923 after the dissolution of the multi-ethnic, multi-religious and multi-lingual Ottoman Empire with the aim of instituting a modern nation state and cultivating a national culture, which in turn would adopt the means of modernity, nationalism and secularism. This imagined nation, which was defined as "Turks," would be comprised of Turkish-speaking, "Muslim born yet secular subjects, who are different from Arabs and keen to adopt western life practices and technologies."[10] Yet the transformation of the decrepit empire into a modern nation-state has been a painful process, especially for the Kurdish population, who had a strong sense of Kurdishness, had largely spoken only Kurdish, and enjoyed autonomy under the rule of the Ottoman Empire where the common bond between Kurds and Turks was Islam.[11] While the policies of the newly founded state were not appreciated by various groups, such as Islamists, in the Republic's early years the most persistent resistance to the modern and secular nation state came from Kurdish-populated areas, particularly against policies which aimed to develop a nationhood based on the Turkish language.[12] As Mesut Yeğen suggests, "Kurdish identity was one of the victims of the political project of building a modern, central and secular nation state, the necessary condition of which was the exclusion of religion, tradition and the periphery."[13] Accordingly, the resistance which emerged out of the Kurdish population was represented as the resistance of pre-modernity and of banditry.[14] Kurds who claimed their ethnic identity were seen not only as a threat to the unity of the nation-state, but also to the modernity ideals that are necessarily exclusive of local fabric, cultural and religious practices.

Turkey's modernization and centralization policies had been successful to a certain extent, especially in repressing resistance coming out of the Kurdish population, but by the late 1960s, a number of leftist groups began to raise the Kurdish question as an issue. With the rise of the Kurdish guerilla movement PKK in the early 1980s, the ethnic and cultural consciousness grew and spread amongst almost all Kurds living in Turkey.[15] Violence and human rights violations against Kurds during the military dictatorship of 1980–83 and prohibition in 1983 of the use of Kurdish in all walks of life, even ordinary speech between individuals, provoked young Kurds especially to become involved in activism. Some were even motivated to join guerilla forces, even though the average life of a guerilla in the mountains of the Kurdish area was known to be just three years.[16] Although the Turkish state eased the ban on Kurdish in the early 1990s, the public use of Kurdish in education, the judiciary and legislation is still prohibited. As a result, many young Kurds speak Turkish as their first language or use only Turkish as their written language even in the southeastern part of Turkey where predominantly Kurds live. Thus, they have been dissociated from a number of things: their right to speak and write in their maternal language, their right to claim their cultural and ethnic identity, their right to mourn for their losses and recall their common memory.

For millions of politically active Kurds, the use of media technologies became of crucial significance, signifying the possibility of expressing themselves, making and inventing themselves in the way they imagine, and producing a sense of nation without an officially recognized territory of nation. In the 1990s, with the financial support of the PKK and diaspora Kurds in Europe, the Kurds of Turkey were introduced to the Kurdish television channel MED-TV, a satellite station based in London and Belgium, which made them "the first stateless 'television nation.'"[17] The Kurdish diaspora has also appropriated desktop and electronic publishing to develop a standardized written Kırmanche, the dialect most widely spoken by Kurds in Turkey, as well as an archive of extensive literature, both of which aim to produce a sense of belonging to a culture and a nation with its own literature and history and an institutionalized written language.[18] Very recently, Facebook in Kırmanche has been introduced and found thousands of users immediately, integrating them to the modern and global technoscape. There are signs that a Kurdish cinema is also developing as the voice of pain and agony of millions of people who were deported from their villages and were arrested and tortured and even killed. Likewise, Kurdish music has started to be performed in some cafes or bars in İstanbul and in Diyarbakır. Thus, in contemporary Turkey, Kurds are trying to mourn for their losses, build their self-confidence and assert their cultural and ethnic identity more strongly than ever through creative industry and media technologies.

In such an environment where technologies and cultural productions are attempted to be used for the political ends of the Kurdish population, where does the cell phone stand through its sonic, textual and visual containment and its functions enabling the connection among distant and mobile users? As a fluid technology that allows for the containment of any language, how does the cell phone serve for the young Kurds who have been interdicted from their mother language?

Method

The data are derived from in-depth interviews conducted with 15 cell phone users in Diyarbakır and in İstanbul in April 2011. The sample focused more on the lower middle-class self-defined Kurds and less on the middle class. The reason for the concentration on lower middle-class young adults is that Kurdish-populated areas are largely poorer regions of Turkey and the cell phone is mostly used by young adults, who purchase it through their pocket money or salary. Most of the informants were between 18 and 30 years of age, largely pursuing their university education, with some in employment or seeking work. Eight of the sample were women.

The in-depth interviews, each of which lasted an average of one to two hours, were organized with the aim of learning the meaning and the place of cell phone in the ethnic and cultural identity politics of users. Since defining oneself as Kurd is already a political statement regardless of the users' political engagement with a political party or organization, all of my informants were politically and ethnically conscious subjects. Especially for those living in Diyarbakır, not participating in political discussions or being part of civic resistances against restrictions on the free exercise of "Kurdish identity" is almost out of question, precisely because this city has historically been the "most directly affected by the violence and repression resulting from the confrontation between the PKK and security forces,"[19] housing a large number of internal refugees from the countryside who had been displaced from their homes.

While my research sample is composed of Kurdish users, Kurds do not consist of a unified cultural or linguistic group. For instance, some of my informants who speak a different dialect of Kurdish defined themselves as Kurd first and then as Zaza. Out of the many distinct Kurdish dialects, Kırmanche is the most dominant and institutionalized, while Zaza remains a minority language in Kurdish-populated areas. Young Zaza users were also concerned about the future of their dialect, often complaining that while Kurdish (which almost always refers to Kırmanche in their statements) is repressed by Turkish, Zaza is repressed by both Kurdish and Turkish.

I usually began the interviews with basic questions such as "How do you use your cell or for what purposes do you use it?," to which I received answers that illuminate the significance of using their own language in the ritualized technological practices within their political and cultural struggle. Almost all my informants (except five whose parents were fluent in Turkish) had a traumatic experience of early school years where they were forcefully taught reading and writing in Turkish, which they had no knowledge of. Most of them also recounted incidents of being humiliated as children by Turkish teachers and also other Kurdish students because they could not speak Turkish properly or spoke with a Kurdish accent. Thus, each of my informants had a historically constructed image of Turkish as a language of education, modernity and wealth and of Kurdish as one of tribes, peasantry and illegality. In this respect, I deliberately paid sustained attention to the ways in which technology, modernity, ethnic identity and the forbidden mother language are negotiated in their statements about how they see the place of cell phone in their daily lives and its meaning for their political and social struggles.

Failed Attempt to Patch a Kurdish Menu

In 2009, a local cell phone shop owner, Uğur Akar, collaborated in Diyarbakır with a Turkish distributor of TTN Mobile (a Chinese company) to introduce the first cell phones with Kurdish menus to the local market. These phones allowed users to switch from Kurdish to Turkish whenever they needed to translate a Turkish word into written Kurdish. In the first month of its release, the 500 cell phones with Kurdish menus sold out and new ones were ordered not only by Diyarbakır residents, but also by Kurds from other cities and countries. Sales reached 20,000 in total. Akar was interviewed by newspapers about his success and said, "this is an answer from our side to those who claim that Kurdish is not an everyday language....Kurdish is a language of education, and of technology."[20] In my visit to Diyarbakır in April 2011, I was able to locate neither Akar nor any cell phones with Kurdish menus on the market. Eventually, one of my sources disclosed that Akar had been forced to quit his business under threats and opted to disappear from public.

While the use of Kurdish in the state-run television channel, local newspapers, social media such as Facebook and in some public spaces such as restaurants, cafes, and shops in Kurdish populated areas is no longer subject to severe restrictions, authorities take a more repressive attitude toward the use of Kurdish in cell phones. This actually gives us a sense of the meaning of the cell phone and its place in the landscape of Turkey, including Kurdish-populated areas, in that it is fully embedded in everyday life practices; if the language of the other was incorporated, the state could not deny its existence as a legitimate and common ground for many to communicate and express themselves in daily lives. Süleyman, who developed the first Kurdish software for Linux, comments on this by saying, "they would have no chance of ignoring this language as an unknown one, if they had tolerated its official use in cell phone." Kurdish menus on cell phone thus were not allowed in Turkey and none of my informants in the sample could purchase one of Akar's phones, although some had placed orders months ago. Yet, each of the informants found their own way to place the cell phone in their struggle to embed Kurdish in the daily textual practices of mediascape.

Texting and Speaking in Kurdish on Cellular Telephony

At the banal level, texting and phone conversations enabled Kurdish users to maintain continuous contact with others and to coordinate social and political activities.[21] Yet in its culturally and socially specific usage, texting and telephonic interaction become an experimental, play-like, performative and dutiful work,[22] through which users can practice their "own" language in their daily lives. The practice, in this case, implies two interrelated and inseparable meanings. First, users who do not feel comfortable with speaking or writing in Kurdish exercise their language skills, especially through texting. Second, users of the cell phone appropriate this technology of everyday life to bring Kurdish into the habitual and modern practice of communication and social contact, which contributes to their collective struggle to assert their ethnic and cultural identity.

For users who appropriate the cell phone to practice their language skills, texting in particular becomes a popular site where they improvise and improve their writing skills. Since the majority of the users in the sample used Turkish menus, most were having difficulties writing in Kurdish in proper ways. As Dicle says, "one wrong letter would change the whole meaning of a word and a sentence dramatically in Kurdish." Moreover, half of the sample, especially Zaza users, have never learned how to write in their own language. Thus, the language of texting in a Kurdish context is particularly distinct, because it is an improvisation through the unlearned written language, taking the risk of being incoherent in the name of making the language an essential part of a ritualized social practice of communication. A female Zaza university student in Diyarbakır, Roza, speaks of her determination to text in written Zaza, which she has no knowledge of:

> One of my friends and I have decided to text in Zaza. Because it gets on my nerve to use Turkish in my daily life. It takes forever to write a sentence, it is not practical and easy at all, but we nevertheless continue. We are trying to make it a habit for ourselves.

In a similar vein, Leyla, who is a volunteer teacher in a municipal institution in Diyarbakır, recalls how she and her boyfriend, who lives in İstanbul, started flirting through texting in Kırmanche:

> We started texting in Kurdish and I was so impressed by his Kurdish. It sounded so romantic. Though I am not sure if his Kurdish was romantic or Kurdish was already a romantic language…I noticed that he was hesitating to speak on the phone, always preferring to text. Months later, I found out that his spoken Kurdish really sucks. He learned Kurdish in language courses in İstanbul. An assimilated but a sweet guy, it took ages for him to write those SMS to me.

Thus, texting becomes a way for the Kurdish users to practice their language—not because it is a fast, easy or cheap way of connected presence; on the contrary it is slow, time-consuming and as cheap as talking on the phone. Yet, texting allows them to improvise, perform and experiment with their mother language, which remains unfamiliar to most of them in its written form. The use of improvised written Kurdish in SMS becomes a performative and dutiful practice that is assumed to convey the idealized image of self in the most desired language, which has been subjected to severe interdiction. Other than the dutiful incentives that make users embed their mother's language in the cell phone's containing space, the affective incentives such as hatred and jealousy of Turkish as the assimilator's language, or as the dominant language of modernity and technology, and romantic feelings towards the interdicted mother language, which is presumed to give the feeling of being at home in a language, fuse into each other and mark the performance of texting and conversations for Kurdish users. Rezzan, in her late 20s and working for an NGO in Diyarbakır, explains this well:

> Look, this [cell phone] is something that I interact with all the time and I don't wanna see Turkish here. I mean we're trying to change our habits, all of us had learned reading and writing in Turkish. But this must change and is changing....If were a nationalist, I would definitely be a nationalist language-wise. Not only do I wanna use Turkish in my life, but also I adore Kurdish. It is beautiful. Everything is more beautiful in Kurdish, even love is…

This idea of nationalism grounded in love for a community and a language on the one hand and hatred of the other's on the other is common among many young Kurds. Yet the findings of my empirical research also showed that such sentiments are not generalizable to all politically active Kurds and age factor plays a significant role. While the majority of young adult Kurds appropriate their cell phones as a means of transforming the interdicted mother language into the mother language of all, elder ones do not approach the cell phone or language politics in the same way. Süleyman, a primary school teacher in Diyarbakır in his mid 40s, explains:

> These young people lost the ground of political action, the leftist ideals totally disappeared in their politics. They detest Turks, they hate Turkish language…, this is not progressive, on the contrary they are backward…The language is an instrument, nothing more. I speak Kurdish to my mother on the phone or face to face because my Kurdish is good enough for that. But I cannot do political discussions in Kurdish or text in Kurdish….This does not say anything about my political engagement to our struggle…

Thus, the particular engagement of cellular telephony through an attempt to replace the dominant language of daily communication with the Kurdish dialects by these

Kurds is not only a personal and collective statement of their cultural, ethnic and political identity but also a generational one.

Sonic and Visual Containment of Cell Phones

Through its diverse possibilities of sonic containment and of sonic communication, the cell phone can be considered as a sound technology which holds a central feature in popular music and everyday sonic experience. The sonic containment of the cell phone, such as mp3 sound-sharing files, ringtones and ring-back tones, serves users for the personalization, customization and nationalization of global cellular telephony, and hence user meanings attributed to the cell phone as a sound technology are quite diverse, depending on its actual use.[23] In the case study I present here, sonic data that are housed in the containing space of cell phones have a special place in conveying the users' cultural and political identity in public spaces. The audible cell phone also has a crucial role politically motivated users' efforts to transform the sonic environment of the urban space where Turkish still predominates into a fully Kurdish-speaking environment, as well as to alter the private sonic experience of everyday life practices in such a way that the users would enjoy the feeling of being saturated with sonic data expressing Kurdishness. Among all sonic functions of the cell phone, the ringtones seem to be the most popular playground for these efforts.

The personalized and customized ringtone, which defines both the music taste and cultural/political identity of the user, creates a new private space in public where the prevailing soundscape is transformed through the aggressive interference of the tune into a public setting.[24] This is precisely what makes the ringtone an appealing sonic function for Kurds looking to reverse the predominance of Turkish with the sonic dominance of Kurdish melodies, rhythms, lyrics. Ringtones in diverse Kurdish dialects are so popular, especially in Diyarbakır, that most of my informants told me that they are surprised when they hear a Turkish ringtone, just as they are when they hear a Turkish or foreign song coming out of a cafe in the city. While, using the containing space of the cell phone to house written and spoken Kurdish might be difficult for some users, filling the sonic storage space of their cellular telephony with easily downloadable pirate music informing Kurdishness is doable and enjoyable, as personal taste and collective political duty are simultaneously satisfied through the interference to the public sonic space. Transforming the whole sonic environment also provides the sense of the change of the space, the alteration of the meaning of the space, colonizing the space. Özlem speaks of this experience especially in public transportation vehicles in Diyarbakır: "[It] is fully Kurdish with almost all dialects, passengers' phones are ringing in Kurdish, the radio that is on is in Kurdish, so you feel you are in Kurdistan and not in Turkey."

The generational difference in the use of cell phone for political purposes is also apparent in the way young adults differentiate themselves, especially from the teenagers who, in their eyes, are largely under the influence of Turkish pop culture. Gopinath comments that the ringtone can be understood "as a particularly charged site of social conflicts, in which generational, regional/national, ethnic and even gender divisions are apparent."[25] Young Kurds' appropriation of sonic space using cell phones is not just a personal and collective statement against the dominance of Turkish language and popular music, but also a means of differentiating their position from older or younger Kurds, who are either fearful of using the language by all means in every arena, or reluctant to use it to express their ethnic and cultural identity in public places. İclal, a female university student, who herself cannot speak or write in Kurdish, expresses her frustration:

> The other day, I saw a girl on the street, whose cell was ringing like Serdar Ortaç's song....Quite frankly, I can't stand this. This music is not only cheesy pop music but it is also Turkish trash....Turkish pop culture takes these people under its intoxicating control.

The difference between cell phones and other media technologies such as the television, radio and even the Internet (some of Kurdish websites are still in Turkish), which largely house Turkish language, music, cultural products, is that it allows for the containment of all sonic means of Kurdishness. This becomes a cocoon-like containment for users, who like to feel that they are truly at home, in their "own" language, music and melodies. For instance, Leyla, a university student in Diyarbakır, told me how she wakes up to an alarm that rings a Kurdish song every morning, with the pleasure of "starting my day in peace." Others, who often listen to Kurdish music in diverse dialects, speak of the enjoyment of "living our culture that we have been deprived of," as Mahmut says. Painful memories deriving from events actually witnessed or constructed through overheard stories of the past, where even listening to Kurdish music was banned and punished, become motives for cell phone users to create a sonic imaginary, where the different means of Kurdishness are played out.

Although the cell phone was initially a sound technology, it has evolved to become a more visual technology with cell phones containing video or still cameras, and Internet connections that enable television or video broadcasting on the phone. Although most of my informants did not have smartphones with functions such as email, Twitter, or Facebook, due to their economic status, they had still and video cameras on their cell phones. They often used these cameras to capture the public demonstrations plotted in Diyarbakır or in İstanbul. In that sense, they employ the cell phone to act as an individual broadcasting station through which users upload photos of demonstrations from their cell phone to their computer to share those images on their Facebook account with others and email them to dissident news agencies. When I asked my informants what they keep in their cell phones' visual containing space, most said photos as memories of their private lives, their friends and family members, as well as images of public demonstrations and also some images that reveal their political identity, which some used as wallpaper on their cell phone. Hüseyin, who develops visual images to be communicated in the virtual space of cell phones and the net, says,

> the most popular visuals among Kurds are images of little girls with red, yellow and green bandanas [the colors representing Kurdish community in Turkey] who make victory sign with their hands, photos of Apo [Abdullah Öcalan, the imprisoned founder of PKK, who is the leader of the Kurdish movement in many people's eyes], graphics of free Kurdistan, images of some historical public demonstrations, places in Diyarbakır.

Hüseyin explains his motives by stating, "I do it to support the movement, the movement that is civic," and then he adds,

> we used to tend to hack the nationalists' pages or the states' sites on the net, we were more inclined to destroy rather than to construct something. Now we are more tuned to create our own thing, and the Turkish hackers try to hack us.

Conclusion

Although the instrumental and expressive value of cell phones is well established, I suggest that exploring the place and meaning of cell phones in a dissident political culture enables an analysis of the ways in which everyday resistances are played out through mobile technologies. There is a need to elaborate more on how oppressed young people appropriate the cell phone, an essential technology of containment and connection in everyday life practices, for their political ends.

The containment of the cell phone in the hands of young Kurdish users is necessarily political—political in the sense that it expresses the cultural and political identity of its users, it allows the free exercise of the Kurdish language, and that it enables the circulation of decentralized messages calling for more political action and demonstrations. For all of my informants, the cell phone was the most favored technology, not precisely because it enabled instant and constant connection with others but because it allowed the users to house, experiment and hold their imagined practices of Kurdishness in its fluid containing space. In other words, the cell phone has become a political technology in the hands of Kurdish users precisely because its fluid containing functions are "compatible with particular kinds of political relationships," and the particular use of it became "a way of settling an issue in the affairs of a political community," which is the free and ritualized exercise of Kurdishness.[26] The performance of Kurdishness starts with the integration of the language into the containing space of cellular telephony. Although attempts to patch a Kurdish menu to the containing space of cellular telephony have so far failed, politically motivated young Kurds have found their own ways to use the language, through texting and telephonic interaction and through music. Most of them romanticize the free exercise of their mother language as creating a sense of full self-presence, where they would not feel incomplete, lacking the means of expressing themselves fully, but would reside in the house of their "own" language. Post-structuralist theories inform us that language that initiates subjectivity always remains others' language, which cannot fully be appropriated and which will never give the sense of full self-presence. However, as Derrida discusses, the singular case of political interdiction of mother language presents a singularity of a double interdict where one cannot even find an instrument to invent her own language, "speech" and "diction."[27] I consider young Kurds' passionate, performative and political engagement with the unfamiliar maternal language through the cell phone as an everyday practice of struggle to find an instrument to invent their own speech within the interdicted Kurdish languages and dialects. In this battle, hatred of the colonizers' speech, love of the victimized mother language, Kurdocentric ideologies, claims of progress and modernity, and the technologically ritualized practices of communication are gathered together to generate the performance of cell phones and their users. The sonic, visual and textual containment of Kurdishness in the cell phone creates an instrument for young Kurds to alter the sensual environment they actually inhabit to an imagined one where all freely exercise their language and culture.

My research experience with Kurdish cellular telephony shows that examining its political role in users' lives demands a contextualization of this technology as a daily social practice within larger socio-technical and political conditions. While the role of cell phones in the particular organizations of political demonstrations or smart mobs or crowds has been well covered both empirically and theoretically, the ritualized practice of cell phone as political medium in everyday life requires further research. While the uprisings in the Middle East through the power of mobile technologies including the cell phone, which greatly inspired my informants, is an issue for related studies, the seemingly mundane and ordinary engagement with cell phones as political agents on a daily basis should also be of scholarly interest.

Notes

1. Christian Licoppe, "'Connected' Presence: The Emergence of a New Repertoire for Managing Social Relationships in a Changing Communication Technoscape," *Environment and Planning D* 22 (2004): 135.
2. For an overview, see Eileen Green and Carrie Singleton, "Mobile Connections: An Exploration of the Place of Mobile Phones in Friendship Relations," *The Sociological Review* 57 (2009): 125–144; Kim Yoon, "The Making of Neo-Confucian Cyberkids: Representations of Young Mobile Users in South Korea," *New Media & Society* 8 (2006): 753.
3. Richard Ling, *The Mobile Connection: The Cell Phone's Impact on Society* (San Francisco: Elsevier, 2004); Anandam Kavoori and Noah Arceneaux, eds., *The Cell Phone Reader: Essays in Social Transformation* (New York: Peter Lang, 2006); James E. Katz and Mark A. Aakshus, eds., *Perpetual Contact: Mobile Communication, Private Talk, Public Performance* (Cambridge, UK: Cambridge University Press, 2002).
4. Peter Glotz, Stefan Bertschi and Chris Locke, eds., *Thumb Culture: The Meaning of Mobile Phones in Society* (Bielefeld, Germany: Transcript, 2005); Heather Horst and Daniel Miller, *The Cellphone: An Anthropology of Communication* (Oxford: Berg, 2006); Kristof Nyiri, ed., *A Sense of Place: The Global and the Local in Mobile Communication* (Vienna: Passagen Verlag, 2005).
5. For an exceptional study, see Vicente Rafael, "The Cell Phone and the Crowd: Messianic Politics in the Contemporary Philippines," *Public Culture* 15 (2003): 399–425.
6. Green and Singleton, "Mobile Connections: An Exploration of the Place of Mobile Phones in Friendship Relations."
7. Jonathan Sterne, "Bourdieu, Technique and Technology," *Cultural Studies* 17 (2003): 383.
8. Zoe Sofia, "Container Technologies," *Hypatia* 15 (2000): 181.
9. Adrian De Souza e Silva, "From Cyber to Hybrid: Mobile Technologies as Interfaces of Hybrid Spaces," *Space and Culture* 9 (2006): 270.
10. Yael Navaro-Yashin, *Faces of the State: Secularism and Public Life in Turkey* (Princeton: Princeton University Press, 2003), 12.
11. Kemal Kirişçi, *The Future of Turkish Foreign Policy* (Istanbul: Boğaziçi University Press, 2004).
12. Ibid., 274.
13. Mesut Ye en, "The Kurdish Question in Turkish State Discourse," *Journal of Contemporary History* 34 (1999): 555.
14. Ibid., 563.
15. Kirişçi, *The Future of Turkish Foreign Policy*, 67.
16. Bejan Matur, *Da ın Ardına Bakmak* [Looking Beyond the Mountains] (Istanbul: Tima , 2011).
17. Amin Hassanpour, "Satellite Footprints as National Borders: MED-TV and the Extraterritoriality of State Sovereignty," *Journal of Muslim Minority Affairs* 18 (1998): 39.
18. David Romano, "Modern Communications Technology in Ethnic Nationalist Hands: The Case of Kurds," *Canadian Journal of Political Science* 35 (2002): 130.
19. Kirişçi, *The Future of Turkish Foreign Policy*, 283.
20. "Kürtçe Cep Telefonu Çıktı" [Kurdish Cellphones Are Now on the Market], tumgazeteler.com, last modified June 24, 2010, http://www.tumgazeteler.com/?a=4629585.
21. Richard Harper, Leysia Palen and Alex Taylor, eds., *The Inside Text: Social, Cultural and Design Perspectives on SMS* (Dordrecht, The Netherlands: Springer, 2005).

22. Virpi Oksman and Pirjo Rautuaninen, "Extension of the Hand: Children and Teenagers Relationship with the Mobile Phone in Finland," in *Mediating the Human Body: Technology, Communications and Fashion,* ed. Leopaldina Fortunati, James Katz, and Raimonda Riccini (Mahwah, NJ: Lawrence Erlbaum, 2010), 100; Tony Wilson and Florence Thang, "The Hermeneutic Circle of Cellphone Use: Four Universal Moments in a Malaysian Narrative of Continuing Contact," *New Media & Society* 9 (2007): 950; Aphra Kerr, Julian Kücklich and Pat Brereton, "New Media—New Pleasures?," *International Journal of Cultural Studies* 9 (2006): 70.

23. Michael Bull and Les Back, *The Auditory Culture Reader* (Oxford: Berg, 2003); Heikki Uimonen, "'Sorry, Can't Hear You! I'm on a Train': Ringing Tones, Meanings and the Finnish Soundscape," *Popular Music* 23 (2004): 55; Sumanth Gopinath, "Ringtones or the Auditory Logic of Globalization," *First Monday* 10–12 (2005), http://firstmonday.org/issues/issue10_12/gopinath/index.html.

24. Uimonen, "Sorry, Can't Hear You! I'm on a Train," 58.

25. Gopinath, "Ringtones or the Auditory Logic of Globalization."

26. Langdon Winner, *The Whale and the Reactor: A Search for Limits in the Age of High Technology* (Chicago: University of Chicago Press, 1986), 22.

27. Jacques Derrida, *Monolingualism of the Other, or, The Prosthesis of Origin* (Stanford, CA: Stanford University Press, 1998), 60.

Through the Looking Cell Phone Screen

Dreams of Omniscience in an Age of Mobile Augmented Reality

IMAR DE VRIES

Media disrupt, and they do it well. They bring together spaces and times that would not normally meet. They open up hidden worlds, and insert themselves as monitoring beacons that bridge previously unconnected spheres. They transport sounds, thoughts, images and other forms of information between realms that, although maybe entirely dissimilar on a physical level, thereby gain access to a certain shared conceptual or cognitive stratum. Because of this potential to uncover, connect, and transform, media are disruptive in a seductive sense; they *lure* us into thinking that they are all-powerful gateways, that in the midst of everyday life they can act as magical portals to more knowledge, more experience, and more awareness. Some media do this better than others; in McLuhanian terms, especially the "cool" and "light-through" media beckon us to probe what is "behind" them, to investigate their other side.

McLuhan hailed the 1960s electronic age as ushering in a new era of global awareness, one in which cool media such as televisions and early computer networks would immerse us in a world that had the functioning features of a village. At the beginning of the 21st century, that village has been condensed into small yet über-cool boxes we carry in the palm of our hands, and it is not only us who are always potentially aware of what happens in the world, those small boxes themselves have become sentient of their surroundings as well. The portals that we look through in order to see and know more have thus gained characteristics that make them seem all-powerful indeed; the words that typify our present media ecology best, "digital" and "mobile," both exude an air of morphability and uninhibitedness, traits that answer exceptionally well to a desire to have technologies at our disposal that can connect, translate, and communicate all forms of information imaginable, at all times, in all places.

We know this desire from the way it is abundantly expressed in commercial ads and press releases. However, it would be wrong to state that only smart sale techniques have created the network society we live in. There is more to mobile media than just the commercial promise of making things easy; they are the present-day climactic expression of our technological imaginary, a collection of what could be called "necessary fictions" that tell that life can be improved and that new technologies provide the key to do so. Today's mobile media-saturated societies cultivate an undercurrent of thought that has long been around, one which holds that radically heightened connectedness will, by design and by necessity, bring us closer to a utopia of omniscience, omnipresence and complete understanding. Fueled by this desire for unlimited access to others and information, actual technological innovations as well as visions of things to come therefore often stress improved communication opportunities which are thought to enforce a blurring of hierarchies and a global emancipation of minorities.

However, new media are highly charged with political ambiguity: on the one hand they are celebrated for liberating the individual from the grasps of mass media and governmental deception or oppression, but on the other they re-insert a disciplining mechanism by subjecting users to the semi-masked logic of information and communication networks: connect, or die. This chapter will argue that the current dominant form of discourse that reflects both revered communication ideals and problematic assumptions about radical connectedness can not only be found in the now-familiar context of "social media," but also in that of mobile Augmented Reality (AR). My focus will be on the politics of what could be called network daydreams, the agenda-setting stories of utopian hopes for communication and knowledge immediacy, and on how they operate in the nowadays much-hyped field of mobile AR. These stories, which increasingly mesh with those found in popular and science fiction culture, are much more powerful than is generally acknowledged, and we need to start tracing them more in processes of innovation if we are to understand what the future of mobile media might look like.

Taking the Amsterdam-based mobile AR startup Layar as a prime example of how futuristic visions of mobile and networked AR applications can inspire real-world inventions, I will show that the language used to promote these new AR technologies is filled with politically loaded phrases that emphasize sharing, connecting, liberating, and most of all discovering information that "was always there" but has now finally been made visible. I will argue that the flip side of joining the mobile AR revolution is that users are forced to open up to the logic of the network in which every node is part of a ubiquitous web of information. Mobile AR problematizes the notion of data disclosure more than it has been up to now, and it will do so under the guise of bringing us closer to utopia than ever before.

The Recurrent Dream of Omniscience

When seeking a better understanding of present-day myths that tell of reaching communication perfection by improving media technologies, it is useful to apprehend how and when necessary fictions in the past have generated future-oriented mindsets that were willing to act upon utopian daydreams. It is with such media-archaeological analyses that we can grasp how the assumed enlightening power of telecommunication technologies, which in history has accompanied many technological paradigm shifts, remains to shadow and seep through into the consciousness of a diverse field of scholars, fiction writers, scientists, and engineers in the 21st century. Because of the persistent nature of myths of progress, it is possible to recognize some remarkable discursive similarities in the way new media are framed as being able to augment human reality. Moreover, ever since the end of the Belle Epoque, when network technologies had become very much visible in the everyday lives of people in the shape of the telegraph, the telephone, and early forms of radio and television, we can identify an increasing popular presence of now-familiar enthusiastic expectations that new electronic media could be used to enhance the availability and exchange of ideas, improving the quality of life.

As a case in point, when Wikipedia asks us to join them in their commitment to imagine and realize "a world in which every single person on the planet has free access to the sum of all human knowledge,"[1] or when IBM proclaims on its Smarter Planet website that the employment of mobile intelligent technology will ultimately lead to world peace,[2] they are effectively—and sometimes literally—reiterating 1930s positivist ideas that proper education and information dissemination, managed by the newest telecommunication systems, automatically engender a better world. In 1934, for instance, the Belgian information science pioneer Paul Otlet published his *Traité de Documentation*, a voluminous work in which he proposed to create a documentation system that would be "1° universal in its scope; 2° reliable and correct; 3° complete; 4° quick; 5° updated; 6° easy to obtain; 7° compiled in advance and ready to be communicated; 8° accessible to a great number of people."[3] Otlet envisioned that interaction with this system would make use of a form of networked teleconferencing, in which new media such as film, the gramophone, radio, and television would provide communication and information to a far better degree than the encyclopedic book could offer. As he noted, these instruments would make us equal in perfectness and fullness as to "God himself," because the potential of the new "ubiquitous, universal and eternal" media of his time would help advance humanity towards a "divine state" of "being everywhere, seeing everything, hearing everything and knowing everything."[4]

Otlet was not alone in harboring this techno-religious drive. A similar belief in the need for an enlightening world encyclopedia can be found with H. G. Wells, who in 1938 wrote that the creation of a universally accessible and "complete planetary memory for all mankind" would constitute "a real intellectual unification of our race" and ultimately—foreshadowing IBM's credo—present us with "a way to world peace."[5] Contemporaneously, American engineer Vannevar Bush valued the importance of improving information management in much the same way. His famous article "As We May Think," written in 1939 and published in 1945, described a personal networked microfilm system with which scientists could concurrently store and retrieve documents, and construct "associative trails" between those documents that they could then share among one another. As Bush saw it, such a system might even evolve into an all-electric version and directly link to the brain, elevating "man's spirit" by augmenting his "limited memory."[6]

While it could be argued that such similarities in discourse between the 1930s and our time are mostly found on a rhetorical and not on a material level, it is a mistake to think that only realized technological objects tell the true story of media development. Dream machines and popularized necessary fictions play a far more important role in guiding that development than is usually recognized.[7] They persist in the collective imagination and repeatedly pop up when advances in hardware have created new arenas in which to explore what has become possible. In fact, such new technological realities, in turn, refuel the need to make explicit that the dream of bringing universal enlightenment through advancements in technology has a heritage that needs to be carried forward. This is palpably visible for instance in Joseph Licklider, the director of computer research at the Advanced Research Projects Agency (ARPA) in the first half of the 1960s; he credited Vannevar Bush as a great influence on his thinking, and wrote a number of illustrious texts in which he proclaimed that computers would become personal devices, all connected to a galactic resource network and facilitating the meeting of many interacting minds. As a testament to the agenda-setting power of futuristic visions, Licklider acted upon his visionary ideas; together with his successor at ARPA, Robert Taylor, Licklider was the fundamental initiating force behind the creation of ARPANET, the interconnected science community computer network that would later morph into what we now know as the Internet.

Media-archaeological analyses of recurrent discursive building blocks, then, can give us a more historically contextualized and therefore more comprehensive view on how daydreams and myths have shaped and continue to shape the development of present-day and future technologies. The important thing to note here is that these dreams and myths are not just randomly existent; instead they are part of a long tradition of combining engineering, regulatory, corporate, and consumer voices in a strategic deployment of utopian discourse. In this respect, new media should be regarded to be new in their specific and historically conditioned socio-technological manifestation, not in their incessantly attributed power to further mankind. To be sure, this does not mean that on a more fundamental level things have always been business as usual, that nothing really changes, but what does not seem to change much is that new media are initially met with high hopes, before they become mundane and create room for the expectation of the next big thing.

Mobile AR: The Next Big Thing

Looking at today's media landscape, if there is one area where we can distinguish a renewed boom in utopian stories and attempts to bring us closer to a future of ultimate cooperation and understanding, it surely seems to be that of social media. Popular new communication technologies such as Facebook and Twitter have been ascribed with immense emancipative and transformative power, mainly because of the tempting assumption that their platforms offer uninhibited possibilities to express oneself in a globally accessible and expanding social information network. While reality is of course much more complex than that, in many popular, political and commercial debates the role that social media are cast in is bordering on Messianic, making it seem as if they are able to solve any problem arising from social and cultural differences, and can even topple long-lasting dictatorships. The storyline in these debates follows that of many necessary fictions and is therefore both familiar and powerful: connect, engage, share and learn, and creativity, freedom and democracy will automatically ensue.

However, while social media seem to have captured the undivided attention of business people, scholars and netizens alike at the beginning of the 21st century, it is the fact that our communication media have become mobile that truly deserves recognition. Gathering from industry and user accounts, the utopian desire to augment life is fulfilled to the fullest yet by mobile wireless communication devices. They have become very common in a very short time, and are presented as the seemingly logical, natural, and inevitable outcomes of the ideology of improved communication: unlike any other medium before, they let their users act as both senders and receivers of messages, wherever they are, immersing them in a vast and virtually ubiquitous network of interconnectedness. This makes mobile media the perfect candidates upon which to project repeated and Otlet-like fantasies of ultimate information processing systems, able to blend atoms and bits into an omnipresent and data-rich mixture, accessible and manageable on the fly, by anyone, anywhere.

And indeed, when analyzing popular culture, near-future fiction, and corporate discourse, it is hard to miss the notion that the beginning of the 21st century is when that blend of atoms and bits is going to take concrete shape in the form of augmented reality. We have arrived at what anthropologist Tish Shute calls the age of "World 2.0,"[8] in which digital layers of information are wrapped around so-called World 1.0 and are made accessible through all-powerful and pervasive mobile media. Science fiction writer Bruce Sterling, in an interview with Shute, states that AR "is even more of-our-time than 'social media.' AR has arisen directly from modern technical factors that just didn't use to exist. It's made from shiny new parts, and is truly a child of the twenty-teens, a genuine digital native."[9] While Sterling may be right about AR consisting of "shiny new parts" in a technological sense, I would contend that the underlying driving forces that push AR onto the main stage as the next big thing are still largely the same as those of older—some already long forgotten—media. The recurrent techno-religious motive in media history that stresses a transcendent point in time when everyone will be able to conjure up access to any type of information is clearly found in AR discourse as well.

The repeated presence of this motive clearly reveals itself when we look at the various historically contingent manifestations, both fictional and technologically realized, of systems aimed at augmenting the experience of reality. Usually credited with inventing Virtual Reality, Morton L. Heilig for instance built a system in 1962 called the Sensorama, with which a person could physically immerse herself completely in what Heilig dubbed the "Cinema of the Future." Here, the idea was that a 3D, wide-angle view of moving images with added wind, smell, vibrations, and stereo sound would provide a much better, more realistic mediated experience. Computer scientist Ivan Sutherland, successor of Licklider at ARPA in 1964, expanded upon Heilig's work by exploring the potential of a Head-Mounted-Display (HMD) system that projected digitally created material directly onto a viewer's eyes. In his aptly named 1965 essay "The Ultimate Display," Sutherland holds that "[a] display connected to a digital computer...is a looking glass into a mathematical wonderland," and:

> The ultimate display would, of course, be a room within which the computer can control the existence of matter. A chair displayed in such a room would be good enough

to sit in. Handcuffs displayed in such a room would be confining, and a bullet displayed in such a room would be fatal. With appropriate programming such a display could literally be the Wonderland into which Alice walked.[10]

The references to Lewis Caroll's work are not without significance, I would argue; as I stated earlier, this is exactly what new media technologies (and especially augmented reality systems) do—they let themselves be cast in the role of being able to magically grant access to another world, to let us see what was always already there, hidden from view until an appropriate and well-calibrated looking glass could make it appear right before our eyes.

The theme of merging coexistent yet separate realms in order to make information accessible that is there-but-not-yet-there emerges again and again in mixed reality history. With computer hardware becoming more powerful and Sutherland's visions inspiring new generations of researchers working on Human-Computer Interfaces (HCIs), the decades following the first experiments with HMDs sparked newly articulated but familiar necessary fictions about improving man's capabilities by forging symbiotic relationships between physical reality and virtual data space.[11] What is important to note is that we find these fictions not only motivating the growing computer industry, but also inspiring popular culture; William Gibson's novel *Neuromancer* is of course best known for providing the archetypical 1980s depiction of a virtual world that people could jack into at any time—and it is again no fluke that the book's title is derived from the word "necromancer," meaning a magician able to call into ghostly being the spirits of dead organisms. When the 1990s brought the global adoption of the Internet and the World Wide Web, commentators not only declared that Gibson's cyberspace had become a reality and offered us a world in which "all the sentiments and expressions of humanity, from the debasing to the angelic, are parts of a seamless whole, the global conversation of bits,"[12] but also predicted that "increasing mobility and the miniaturization of communication devices" would reduce our virtual world interaction technologies "to a chip implanted in a retina…connected by cellular transmission to the Net," radically augmenting our experience of reality.[13]

More recent AR developments have indeed relied on computer chips and display technologies becoming smaller, more powerful, cheaper, and therefore more ubiquitous; the rationale behind AR improvement, however, has not changed much. In his classic 1997 survey of the field of AR, Ronald Azuma notes that combining real and virtual objects is useful because AR "enhances a user's perception of and interaction with the real world" by displaying virtual objects "that the user cannot directly detect with his own senses."[14] To this he adds, in a manner Vannevar Bush would have undoubtedly endorsed, that AR is a specific example of "Intelligence Amplification." In the 2001 follow-up survey, Azuma et al. refrain from making such explicit statements about the role AR could or should have in lifting man's spirit, but implicitly they reconfirm AR as the next big thing to help us advance when they say that "AR's growth and progress have been remarkable," and that the field "needs an updated survey to guide and encourage further research in this exciting area."[15] What is more, they signal that "through films and television, many people are familiar with images of simulated AR," implying that popular culture has already paved the way for AR to become an accepted part of everyday life, and that the technology needs to catch up with the daydreams.[16]

Today, the AR industry is visibly busy catching up, and it does so via apparatuses that have thoroughly been accepted as integral parts of our lives. When asked by Tish Shute which technologies have shown the most promise for AR, Bruce Sterling answers: "It's got to be handsets. Smartphones. The stats there are just amazing. The smartphone biz makes the personal computer business look like a Victorian railroad."[17] And he is right; mobile communication devices are nowadays often equipped with a camera, a compass, and Global Positioning System (GPS) electronics, all basic ingredients for making mobile location-aware AR possible. They are virtually ubiquitous and they seem to carry an aura of endless potential, which makes them excellent catalysts for innovation dreams. It should therefore be no surprise that within the institutionalized processes of mobile innovation networks we can find the roots of today's largest mobile AR company, Layar. Born out of a series of Mobile Monday meetings in Amsterdam, Layar presents us with the most advanced and widely publicly available form of mobile AR yet, and, accordingly, with some telling examples of how necessary fictions, myths and daydreams have pervaded mobile media's discourse, and guide its development.

Layar: Supplying the Magic

Layar is a subsidiary of SPRXmobile, an Amsterdam-based start-up company founded by Raimo van der Klein, Maarten Lens-FitzGerald and Claire Boonstra (for space-saving reasons, I will refer to them collectively as "KLB"). Its biggest product is the mobile Layar application, with the May 2011 statistics showing it supports 1.5 million active users, 2150 active layers, 15 languages, and 4 major mobile platforms.[18] The application uses GPS or cell phone triangulation to determine the spatial location of the mobile device it is running on, and connects to the Layar database to retrieve information about nearby Points of Interest (POIs) such as ATMs, houses for sale, Twitter messages, etc. It then superimposes digital representations of these POIs onto the camera-provided and real-time images of physical reality. Layar's infrastructure is designed to be open; anyone can download the appropriate tools to create a set of data files that, when approved by Layar's reviewing staff, can be published online and made accessible to the mobile application.

When we look at its origin story, it is clear that right from the start the Layar project was to be infused with an aura of enchantment and endless possibilities, borrowed from familiar discourses found in the growing mobile media industry. With backgrounds ranging from Web 2.0 strategy to mobile marketing and development at Nokia and KPN Mobile, and having initiated the launch of the Dutch edition of the Mobile Monday concept in 2007, KLB decided in 2008 that they came well equipped to create an enterprise that, as they state on the SPRXmobile website, would dedicate itself to "[designing] services around the personal context of your customer in such an intuitive way that it seems like waving a magic wand."[19] The language of sorcery used here clearly reverberates Sutherland's desire to have augmented reality technology enable us to step into a wonderland at will, and it has proven to be a very powerful discursive strategy. Stressing the magic theme within the mobile AR industry has even become something of a norm; Gene Becker, Layar's AR Strategist, regularly talks about "magic lenses" and "magic windows" when describing mobile AR systems, and user experience designer Mike Kuniavsky argues that, within the context of pervasive AR, "magic as a design metaphor" is especially useful to "help users understand how newfangled ubiquitous computing products can be used."[20] The implication of having all that magic at our finger tips, of course, is that we gain in power, are able to take more command over our surroundings, and therefore come closer to realizing the dream of being omnipotent and omniscient.

What is particularly striking about the way in which KLB frame Layar's founding history is that they explicitly link it to the AR-pervaded worlds of *Rainbows End*, the 2006 near-future novel by Vernor Vinge, and of *Dennou Coil*, the 2007 Japanese anime series directed by Mitsuo Iso. In *Rainbows End*, set in the year 2025, people wear contact lenses and smart clothing in order to uninterruptedly immerse themselves in many different AR layers, and in *Dennou Coil* (which literally translates to "Electric Brain Coil"), set in 2026, the protagonists are children who live in a hyper-connected hybrid world where the virtual and the physical have truly blended. Whenever they can, KLB reveal they were greatly inspired by both book and series, with Lens-Fitzgerald adding in an interview on ReadWriteWeb that because of these inspirations "augmented reality was on our list of things that we wanted to do."[21] What we see at work here is the affective power of necessary fictions, as expressed in fantastical near-future daydreams, and their tightly intertwined relationship with actual decision-making processes. More importantly, we are also able to see how these fictions are selectively mined for their attractive and luring properties, strategically de-presenting the problematic political and ethical consequences of making them a reality. In *Rainbows End,* for instance, not all layers are accessible to all users because of a fragmented AR ecology. Also, people gather in what are called "belief circles," groups that support a specific AR layer and oppose members supporting other layers with the underlying motive that their respective layer should be elected the default view. At the end of the book, so many groups are fighting over the supremacy of AR space that the arrival of the ultimate method of convincing people to adopt a definitive viewpoint, one that has been dreaded throughout the novel, is now seen as almost a welcome relief: YGBM (You-Gotta-Believe-Me) technology, a form of mind control. Similar problems with perceptions of reality and conflicting worldviews occur in *Dennou Coil,* where the inhabitants of the partly physical, partly virtual city of Daikoku are constantly challenged on a real-time basis to assess the exact nature of—and remain firmly planted in—their environment: essential infrastructural elements remain invisible without the right AR tools, and even if one has access to all virtual layers, there is the danger of becoming so much engulfed by them that it induces a Virtual Reality coma.

The problematic ontological relationship that people in *Rainbows End* and *Dennou Coil* have with their surroundings clearly recalls the classical debate about whether new media by definition make us lose sight of our presumed "authentic" reality; what the book and series add, however, is a critical angle to the idea that a radical blending of physical and virtual information is *necessarily* an improvement on how we gain knowledge of and interact with the world. In Layar's AR discourse, however, this blending is presented as the next natural step in dealing with reality per se. With a deliberate reference to how earlier media have developed to come to define our everyday life, and thereby unambiguously invoking the power of myths of information technology's progress, Gene Becker for instance claims that Layar is "still in the silent film era of the AR medium."[22] KLB, as well, state that "augmented reality is a new mass medium" and that "the medium is really in its infancy, we're only getting started. We're in the 1994 of the Internet, we're at the moment when radio became tv."[23] With mobile AR, it seems, enriched access to any type of information at any time and at any location will soon be a reality, and in Layar's view that reality will contain none of the problematic issues raised in *Rainbows End* and *Dennou Coil.*

It would of course be a stretch to say that Layar is after our thoughts or that too much playing around with their app will result in people not being able to tell physical reality and VR apart. However, these hypothetical scenarios can be considered as radically extrapolated versions of real ills that plague AR ideology. From reading that ideology as the recurrent and present-day rendering of Otlet's dream of omniscience, after all, it follows that the desire to increasingly know more and have access to a myriad of information necessarily implies that all kinds of data should be networked, that those networks should converge, and that their contents should be translated into an easily accessible and common readable language, up to the point where there is no ontological difference between data entries and everyone is able to instantly know what anyone else knows. Mobile AR at present is a long way from reaching that zenith, but that does not stop KLB from proclaiming it will do so eventually, and that Layar presents the first step toward that goal. In their 2011 company vision and philosophy video, they employ the type of discourse that is wonderfully—magically—suited for this claim:

> Whatever you want, whatever you want to see, it's all in there....One of the best things about augmented reality is that you can create anything, you can make your own world....There are no barriers, there are no limits to what is possible when your reality is your canvas....Every day at Layar we get one step closer to realizing our dream, and every day brings more confidence that we are on the right track.[24]

Clearly at play here is a combination of willful amnesia, strategic deployment of recurrent discursive building blocks, and hopeful technological imaginary. What is crucial to understand is that these words are not just innocent fantasies straitjacketed in marketing mumbo jumbo; they *do* something, they speak a familiar language of progress with which they aim to reconfirm that the path towards true omniscience and omnipotence actually exists—and that it runs through mobile augmented fields. The only thing we have to do to follow that path to its sublime end point, it seems, is to accept the inherent logic of a radically pervasive knowledge space, which is that everything should be connected to everything else.

Leaving aside the more futuristic problems of unified minds and comatose VR prisoners, a forthcoming division of people into those that do and those that do not have access to an augmented worldview is very much on the cards. Layar's October 2010 press release on what it called its "uninvited DIY exhibition" makes no qualms about this:

> On Saturday October 9th, the physical space inside the MoMA building in New York will host a virtual exhibition, based on Augmented Reality technology. The show will not be visible to regular visitors of the MoMA, but those using the "Layar Augmented Reality browser" on their iPhone or Android smartphones will see numerous additional works on each of the floors.[25]

The marketing motive here is obvious, people should use Layar's app and not anyone else's. As a consequence, however, the visibility of virtual data could increasingly become dependent on using its distributor's designated platform. KLB are well aware that they need to obscure this image of a fragmented future of mobile AR, so in their pronouncements they hark back to ready-at-hand and oft-repeated human advancements that new media allegedly offer, such as equal opportunities for all and liberation from oppression. In a January 2011 press release, Boonstra stated that "this year is about the democratization of augmented reality as we work to find ways to make it easier to create and publish AR content for all,"[26] and a few months later Lens-Fitzgerald contended that "augmented reality will liberate us al [*sic*] from thos [*sic*] who control places and objects. It will be an equalizer."[27] They stop short of foreseeing the advent of world peace, but their message is clear: when everyone has a magic wand to augment reality, things can only get better.

A Popular Culture's Future?

As the present-day and arguably most advanced mass-produced technological manifestations of what the quest for omniscience, omnipresence, and omnipotence is able to produce, mobile communication media have cemented their place in our everyday lives. Their luring power of potentiality, perceived as the result of many incremental answers to long-held dreams of communication improvement, has nowadays reached a stage where it can be directed to see and know more of our direct physical surroundings. In mobile AR, then, the recurrent myths of progress have found a new platform, and they continue to produce the type of discourse that reflects the inherent transcendental nature of an imagined sublime endpoint. Companies that create AR content call themselves *Total Immersion* or *Beyond Reality*, advertisements tell that AR is "reality, only better"[28] or "reality reinvented,"[29] and when looking at the people at Layar it is clear that what drives them is the desire to make real a world in which having magic-like powers will be a possibility for everyone.

Implicit in mobile AR discourse is the fundamental precondition for such a magical world to exist, which is that everything must be radically connected in a vast information network. Atoms and bits can only blend if they become part of a single universe, a feat that will require a staggering amount of data gathering, storing, and disclosure. Unsurprisingly, then, this is precisely what the AR industry is pushing for when they ask us to join in, discover, co-create, and share. Only when users yield to the logic of the all-encompassing network will the experience of AR indeed be as miraculous and seamless as portrayed in popular fiction such as *Rainbows End* or *Dennou Coil*. But it will also be as problematic, and without critical analyses of how problems are de-presented, we run the risk of entering a mobile AR environment that is far removed from actual day-to-day life experiences.

Layar is fast on its way to becoming synonymous with mobile AR, so when we magically jump through the looking cellphone screen, we might do so out of a desire to go beyond the already-known, to experience a wondrous world where everything seems possible, but it could be Layar's view of that world for some time to come. Only if AR protocols become standardized in some way might something like AR neutrality appear, at least in a technical sense. Until then, people could face the prospect of having no other choice than to accept the de-politicized interpretations that Layar makes of near-future AR scenarios. When those interpretations contain musings on how a movie like *Blade Runner* can inspire the company to focus on the role of the eye in future mobile AR technology,[30] or how the pending arrival of Terminator-like "bionic contact lenses" has already brought that future forward,[31] it is hard not to conclude that Layar, as the de facto market leader, enforces as default a mobile AR ecology that is firmly rooted in popular culture imagery but devoid of pesky problems, making it all the more tempting to go along with it. Studying Layar is watching the full force of necessary fictions in action.

Notes

1. "Home," *Wikimedia,* accessed April 19, 2011, http://wikimediafoundation.org.
2. "Smarter Goverment," IBM, accessed April 19, 2011, http://www.ibm.com/smarter-planet/us/en/government/ideas.
3. Paul Otlet, *Traité de Documentation: Le Livre sur le Livre, Théorie et Pratique* (Brussels: Editiones Mundaneum, 1934), 6. All translations from Otlet's work are the author's.
4. Ibid., 431.
5. H. G. Wells, *World Brain* (Garden City: Doubleday, Doran & Co., 1938), 86, 88.
6. Vannevar Bush, "As We May Think," *The Atlantic Monthly* 176, no. 1 (1945), accessed April 19, 2011, http://www.theatlantic.com/doc/194507/bush.
7. Cf. Eric Kluitenberg, *Book of Imaginary Media: Excavating the Dream of the Ultimate Communication Medium* (Rotterdam, The Netherlands: NAi, 2007).
8. Tish Shute, "Home," *UgoTrade* (blog), accessed April 19, 2011, http://www.ugotrade.com.
9. Tish Shute, "Augmented Reality—Transitioning Out of the Old-Fashioned 'Legacy Internet': Interview with Bruce Sterling," *UgoTrade* (blog), May 6, 2011, accessed May 9, 2011, http://www.ugotrade.com/2011/05/06/augmented-reality-transitioning-out-of-the-old-fashioned-legacy-internet-interview-with-bruce-sterling.
10. Ivan Sutherland, "The Ultimate Display," *Proceedings of IFIP Congress* (1965), 506, 508.
11. Most notably in Scott S. Fisher's work; see his publications on dataspace, telepresence and virtual environments at http://www.itofisher.com/PEOPLE/sfisher/publications.html.
12. John Perry Barlow, "A Declaration of the Independence of Cyberspace," *Barlow Home(stead)Page,* accessed April 19, 2011, https://projects.eff.org/~barlow/Declaration-Final.html.
13. Lev Manovich, "An Archeology of a Computer Screen," Lev Manovich: Articles, accessed April 19, 20011, http://www.manovich.net/TEXT/digital_nature.html.
14. Ronald T. Azuma, "A Survey of Augmented Reality," Ronald T. Azuma: My Publications, accessed April 19, 2011, http://www.cs.unc.edu/~azuma/ARpresence.pdf, p. 3.
15. Ronald T. Azuma et al., "Recent Advances in Augmented Reality," *IEEE Computer Graphics and Applications* 21, no. 6 (2001): 34.

16. Ibid., 43–44.
17. Shute, "Augmented Reality."
18. "Layar Sales Support Information for Developers," Layar Development Support, accessed May 9, 2011, http://devsupport.layar.com/attachments/token/4ad17ab8uyc03zr/?name =Layar_sales_support_for_developers.pdf, p. 26.
19. "Home," SPRXMobile Mobile Service Architects, accessed April 19, 2011, http://www.sprxmobile.com.
20. Mike Kuniavsky, "Magic as a Metaphor for Ubiquitous Computing," *Ambidextrous* 6 (2007): 36, accessed May 9, 2011, http://www.orangecone.com/ambidextrous_i6p36 _37.pdf.
21. Mike Melanson, "2WAY Q&A: Layar's Maarten Lens-FitzGerald on Building a Digital Layer on Top of the World," *ReadWriteWeb* (blog), May 18, 2011, accessed May 19, 2011, http://www.readwriteweb.com/archives/2way_qa_layars_maarten_lens-fitzgerald_on_building_a_digital_layer_on_top_of_the_world.php.
22. Gene Becker, "Developing Mobile AR with Layar," Slideshare, accessed May 20, 2011, http://www.slideshare.net/layarmobile/layar-gene-becker-at-are2011, slide 28.
23. Layarmobile, "Layar—Impactful Augmented Reality in Your Everyday Life," *YouTube*, accessed April 19, 2011, http://www.youtube.com/watch?v=HW9gU_4AUCA.
24. Ibid.
25. "Uninvited DIY Exhibition at MoMA NYC," Layar blog, April 8, 2010, accessed April 19, 2011, http://site.layar.com/company/blog/uninvited-diy-exhibition-at-moma-nyc.
26. Chris Cameron, "Layar's Augmented Reality Now Made Possible for All iPhone Apps," Layar blog, January 27, 2011, accessed May 9, 2011, http://site.layar.com/company/ blog/layars-augmented-reality-now-made-possible-for-all-iphone-apps.
27. Dutchcowboy, "@retailigence Augmented reality will liberate us al from thos who control places and objects. It will be an equalizer," Twitter post, accessed May 9, 2011, http://twit-ter.com/#!/Dutchcowboy/status/66484761878724608.
28. TechnologyforGood, "Augmented Reality at a Restaurant," *YouTube*, accessed April 19, 2011, http://www.youtube.com/watch?v=KIwcYHRbgY4.
29. "Home," Ogmento, accessed April 19, 2011, http://ogmento.com.
30. Raimo van der Klein, "YouTube Goodness: Week 1—The Eyes of Blade Runner," Layar blog, May 13, 2011, accessed May 20, 2011, http://site.layar.com/company/blog/youtube-goodness-week-1-the-eyes-of-blade-runner.
31. Chris Cameron, "Watson, Augmented Reality and the Future," Layar blog, February 2, 2011, accessed April 19, 2011, http://site.layar.com/company/blog/watson-augmented-real-ity-and-the-future.

Contributors

Editors

Noah Arceneaux is an Assistant Professor in the School of Journalism and Media Studies at San Diego State University. His research explores the social construction of new media technologies, ranging from wireless telegraphy to emerging forms of mobile media. His work has been published in the *Journal of Broadcasting and Electronic Media, New Media & Society, Technology & Culture,* and *Journalism and Mass Communication Quarterly.* Prior to joining academia, he produced websites for the ABC, CBS, and Fox television networks.

Anandam Kavoori is a Professor at the Grady College of Journalism and Mass Communication, University of Georgia. He is the author or editor of eight scholarly books and over 40 journal articles and book chapters. Amongst his recent book publications are *Digital Media Criticism* (Peter Lang, 2010), *The Logics of Globalization* (Rowman & Littlefield, 2009), and *Global Bollywood* (New York University Press, 2008). He has published scholarly articles in many leading international journals including *Media, Culture and Society, Journal of Communication, Global Media Journal, Critical Studies in Media Communication, Communication Monographs, Journal of Broadcasting and Electronic Media, Journal of International Communication,* and *International Journal of Cultural Studies.* He is also the author of two works of fiction.

Contributors

Ben Aslinger is an Assistant Professor in the Department of English and Media Studies at Bentley University (Waltham, MA). His research has been published in the journals *Popular Communication: The International Journal of Media and Culture* and *The Velvet Light Trap: A Critical Journal of Film and Television* as well as the anthologies *Down to Earth: Satellite Technologies, Industries, and Cultures, LGBT Identity and Online New Media,* and *Teen Television: Essays on Programming and Fandom.* His current projects include a book on popular music licensing in television and video games, and *Gaming Globally: Production, Play, and Place,* a collection on gaming and globalization co-edited with Nina Huntemann (Suffolk University).

Burçe Çelik is an Assistant Professor in the Department of New Media at Bahçeşehir University, Istanbul. Her research focuses on cultural studies of technology, history of communication technologies and cellular telephony. She is the author of *Technology and National Identity: Mobile Communications and the Evolution of Post-Ottoman Nation* (I&B Tauris, 2011). She also holds a doctorate in Communications from McGill University.

Aymar Jean Christian is a doctoral candidate in the Annenberg School for Communication at the University of Pennsylvania. His research focuses on the cultural production of new media, particularly web video and television. He has published in *Cinema Journal, Communication Culture & Critique, Transformative Works & Culture* and *First Monday.* He writes regularly for his own blog, *Televisual,* and numerous print and web publications.

Imar de Vries is an Assistant Professor at the Department of Media and Culture Studies at Utrecht University. His research focuses on media archaeology, mobile augmented reality, and technological imaginaries in popular culture. His forthcoming book *Tantalisingly Close,* on communication desires in mobile wireless media, will be published in 2012 by Amsterdam University Press.

Jason Farman is an Assistant Professor at University of Maryland, College Park in the Department of American Studies and a Distinguished Faculty Fellow in the Digital Cultures and Creativity Program. He is author of *Mobile Interface Theory: Embodied Space and Locative Media* (Routledge, 2011). He received his Ph.D. in Digital Media and Performance Studies from the University of California, Los Angeles.

Gerard Goggin is Professor of Media and Communications in the Department of Media and Communications, the University of Sydney. His books include *Mobile*

Technology and Place (Routledge, 2012), *Global Mobile Media* (Routledge, 2011), *New Technologies and the Media* (Palgrave, 2012), *Cell Phone Culture* (2006), and *Digital Disability* (Rowman & Littlefield, 2003). He holds a doctorate in English Literature from the University of Sydney.

Caroline Hamilton is Australian Endeavour Fellow researching in collaboration with the Institute for the Future of the Book and McKenzie Research Fellow in Publishing and Communications at the University of Melbourne. She is the author of *One Man Zeitgeist: Dave Eggers, Publishing and Publicity* (Continuum, 2010). She holds a doctorate in English Literature from the University of Sydney.

Thomas W. Hazlett is Professor of Law and Economics at George Mason University, where he also serves as Director of the Information Economy Project. He previously taught at the University of California, Davis and the Wharton School, and is a columnist for the *Financial Times*. Prof. Hazlett has also written for the *Wall Street Journal, New York Times, Slate, Barron's, The Economist, The New Republic,* and *The Weekly Standard*, while publishing academic research in the *RAND Journal of Economics, Journal of Law & Economics, Journal of Financial Economics, Journal of Economic Perspectives, University of Pennsylvania Law Review* and the *Columbia Law Review*. In 1991–92 he served as Chief Economist of the Federal Communications Commission.

Matthew A. Killmeier is Associate Professor of Communication and Media Studies at the University of Southern Maine. His research primarily concerns American radio history, media history and media ecology. He has recently published articles on the body as a communication medium, collective memories of World War II in the Japanese press (with Naomi Chiba), the use of music in TV political advertisements (with Paul Christiansen), the radio adaptation of *The Third Man* and the horror radio program *The Hall of Fantasy*. Currently he is working on a book on 1930s–50s horror/mystery radio programs. Dr. Killmeier received his Ph.D. from the University of Iowa in 2003.

Scott W. Ruston is currently an Assistant Research Professor with Arizona State University's Hugh Downs School of Human Communication, where he specializes in narrative theory and media studies. Dr. Ruston's research addresses the art, education and entertainment uses of mobile and interactive media. He has published his work in journals such as *Storyworlds: A Journal of Narrative Studies* and *The International Journal of Technology and Human Interaction*, and he has served as a consultant for mobile entertainment companies and trade groups. Further addressing narrative theory, media technologies and socio-cultural phenomena, he is co-author of *Narrative Landmines: Rumors, Islamist Extremism and the Struggle for Strategic Influence* (Rutgers University Press, 2012).

Collette Snowden is a Program Director of the Communication and Media Management Program at the University of South Australia, School of Communication, International Studies and Languages. She was the inaugural Donald Dyer Research Scholar at the University of South Australia. Her work is informed by more than 25 years of professional experience as a media professional, in various positions including as a print and broadcast journalist, community media activist, political media strategist, and in senior roles in public affairs management and public relations.

Samuel Tobin is an Assistant Professor of Communications Media at Fitchburg State College in Massachusetts who works on play, media and everyday life. His work has appeared in *Games and Culture, The Journal of Popular Noise* and the book *Pikachu's Global Adventure* (Duke University Press, 2004). Currently he is completing a manuscript on mobile game play practices. He holds a doctorate in Sociology from the New School for Social Research in New York City.

Index

Advanced Research Projects Agency (ARPA), 180

affordances, of mobile media, 23–39

Agar, Jon, 16

Anderson, Chris, 87

Android (Droid), 31–32, 36, 79, 81, 89–90, 92, 94–95, 113–114, 159, 187

Apple iOS, 89

Arceneaux, Noah, 1–8, 55–68, 191

ASCAP, 156

Aslinger, Ben, 7, 148–162, 192

AT & T, 26, 30, 37, 69, 71, 75, 78, 80

AT & T Mobile TV, 26, 37

augmented reality (AR), 19, 177–190

Australian Broadcasting Commission, 127

auto radio, 13, 40–54

batteries, 15

Bell Laboratories (AT & T), 16, 69

Benjamin, Walter, 140

BlackBerry, 92, 96

BlipTV, 88

Boxing Day tsunami, 125

Bull, Michael, 13

Buñuel, Luis, 26, 32

Bush, Vannevar, 109, 180

Carey, James, 40–41

Çelik, Burçe, 7–8, 163–176, 192

cell phone, origins of, 15–17, 69–70, 72

cell phone novels, 107–109

China, 9, 109, 124

    cell phone novels, 109

    control of information, 124

Christchurch earthquake, 127

Christian, Aymar Jean, 6, 87–101, 192

Cingular, 78

citizen journalists, 123–124

citizens band radio (CB), 13–15, 55–68

    jargon, 58

codex, 104

"Convoy" (song), 59

Convoy (film), 60
Crackle, 88, 97
Crazy Frog, 156

de Certeau, Michel, 57
de Vries, Imar, 8, 149, 177–190
Derrida, Jacques, 174
Donner, Jonathan, 16
drive time, 45, 47
Droid. *See* Android

e-readers, 102–119
Egypt (ancient), 11
Ericsson, Lars Magnus, 15
ESPN Mobile, 28

Facebook, 25, 34, 37, 94, 166, 168, 172–173, 181
Farman, Jason, 3–4, 9–22, 192
FCC, 57–58, 60, 64–65, 69–83, 87, 90, 97
    creation of, 70
Flash, 90–93
Fleet Call, 74
FloTV (MediaFLO), 26, 37
flow, concept of, 49–50
Ford, Betty, 59
Fortunati, Leopoldina, 107
Foursquare, 14, 34, 36
Friendster, 19
Funny or Die, 88

Gates, Bill, 110
geocaching, 19
geotagging, 19, 56
Gergen, Kenneth J., 17–18
Gitelman, Lisa, 10–11, 20
Goffman, Erving, 136–137, 139, 144
Goggin, Gerard, 6, 31, 55–56, 102–119, 192–193
Google (*See also* Android), 79, 81, 89–90, 95–97, 113–114
Gowalla, 14
GPS, 19, 184
Gunning, Tom, 23, 26

Gutenberg, Johannes, 12, 106

Hamilton, Caroline, 6, 102–119, 193
Hamilton, Shane, 61
Hardt, Hanno, 127
Hayles, N. Katherine, 11
Hazlett, Thomas W., 5, 69–86, 193
hieroglyphics, 11
Hilton, Paris, 31
Hindenburg disaster, 129
Hjorth, Larissa, 108
homing pigeons, 121
homosexuality, 64
HTML 5, 90–93, 96
Huffington, Arianna, 124
Huffington Post, 128
Hulu, 88, 96

i-Mode, 108, 149
Innis, Harold, 11, 135
iPad, 9, 90, 92, 97, 103, 114–116
iPhone, 9, 21, 23, 36, 81, 90, 92, 95, 98, 103, 112–115, 137, 187
iPod, 9, 13, 114, 158
iTunes, 148, 151

Japan, cell phone novels, 107–109

Katz, James, 28–29
Kavoori, Anandam, 1–8, 191
Kay, Alan, 110
Killmeier, Matthew A., 4–5, 40–54, 193
Kindle, 103, 113
Kobo, 103, 113
Kurdish television, 166

Layar, 178
Licklider, Joseph, 180
Licoppe, Christian, 149
Ling, Rich, 16, 28–29,
location-based services, 19–21, 34–36, 56, 184
London Underground, bombing, 126
Loopt, 14

Maho i-Land, 108
Matrix, The, 35
McLuhan, Marshall, 177
Microsoft Windows Phone, 29
mobile gaming, 135–147
mobile media, 9–21, 23–27, 87–101,
  148–162
  cinematic representations, 23–37
  e-readers, 102–119
  gaming, 135–147
  origins of, 9–21
  ringtones, 148–162, 171–172
  spectrum allocation, 69–86
  television, 25–26, 37, 87–101
  use by Kurds, 163–176
mobile privatization, 48
mobile social software (MoSoSo), 55–56
mobile television, 25–26, 37, 87–101
Mobipocket, 111
MobiTV, 97
Motorola, 17, 31–32
Motorola DynaTAC 8000, 17, 29–30
[*murmur*], 20, 36–37
My Damn Channel, 88, 90, 93–96
MySpace, 19
Mytown (app), 36

Netflix, 90, 96
Neuromancer, 183
Newman, James, 137
news ticker, 129
Nintendo DS, 135–147
Nokia, 112, 115
NTT DoCoMo, 108, 149

Olsson, Jan, 23
Open Mobile Video Coalition, 90
Otlet, Paul, 179–180

Pandora, 157–158
papyrus, 11
Pavlik, John, 127
PCS, 74–77
police radio scanners, 129

prostitution, 62–63

Q3030 Network, 88, 93–96
Qualcomm (*See also* FloTV), 26, 152, 74

radio networks, 42–43
railroads, 13
Reformation, 12
Rheingold, Howard, 17
ringtones, 148–162, 171–172
  Billboard chart, 151
  hip hop, 153–154
  Kurdish music, 171–172
  public nuisance, 154–155
Ruston, Scott W., 4, 23–39, 193

Sawchuck, Kim, 26–27,
Shazam, 158
Smokey and the Bandit, 59, 62–63
SMS, 31, 36, 107–109, 151, 164, 170
  sex scandals, 31
Snowden, Collette, 6–7, 120–134, 194
social networks (*See also* Facebook and
  MySpace), 19, 55–56, 65, 123
Sony PortaPak, 125
spectrum allocation, 69–86
  CB radio, 57–58, 64–65
Spotify, 157–158
Sprint, 74, 78, 89–90, 97, 150–151, 153
Stanley Cup 2011, 130
Star Trek, 30
Sterling, Bruce, 182
Sterne, Jonathan, 40–41
Storify, 127
suburbs, growth of, 43–44

T-Mobile, 78, 150
Taylor, Robert, 180
technological determinism, 10
telegraph, 13, 18, 40–41, 66, 128
texting, 18, 107, 169–171
  beginnings of, 107
  Kurdish language, 169–171
Thompson, Clive, 14

Thoreau, Henry David, 18
time zones, 13
Tobin, Samuel, 7, 135–147, 194
transistor radio, 13, 45
Turkle, Sherry, 18
Twitter, 66, 75, 124, 127, 172, 181, 184

Urbanspoon, 34
Uricchio, William, 20–21

VCAST (Verizon), 26, 37, 97, 150
Verizon, 26–27, 89–90, 151
Viacom, 89
Vimeo, 88, 90–93

War of the Worlds (radio broadcast), 129
watches, 12–13
Wells, H.G., 180
Wiebe, Robert, 41
Wikipedia, 179
Williams, Raymond, 48
Winston, Brian, 131
Wu, Tim, 87–88, 97

Yellow Arrow, 36–37
YouTube, 26, 88, 92, 95–97, 126, 158

Digital Formations